Gabriele Zini · Paolo Tartarini

Solar Hydrogen Energy Systems

Science and Technology for the Hydrogen Economy

 Springer

Gabriele Zini
Dipartimento di Ingegneria
Meccanica e Civile
Università di Modena e Reggio Emilia

Paolo Tartarini
Dipartimento di Ingegneria
Meccanica e Civile
Università di Modena e Reggio Emilia

Translated from the original Italian manuscript by Pei-Shu Wu

ISBN 978-88-470-1997-3   ISBN 978-88-470-1998-0 (eBook)
DOI 10.1007/978-88-470-1998-0

Library of Congress Control Number: 2011940674

Springer Milan Heidelberg New York Dordrecht London

© Springer-Verlag Italia 2012

This work is subject to copyright. All rights are reserved by the Publisher, whether the whole or part of the material is concerned, specifically the rights of translation, reprinting, reuse of illustrations, recitation, broadcasting, reproduction on microfilms or in any other physical way, and transmission or information storage and retrieval, electronic adaptation, computer software, or by similar or dissimilar methodology now known or hereafter developed. Exempted from this legal reservation are brief excerpts in connection with reviews or scholarly analysis or material supplied specifically for the purpose of being entered and executed on a computer system, for exclusive use by the purchaser of the work. Duplication of this publication or parts thereof is permitted only under the provisions of the Copyright Law of the Publisher's location, in its current version, and permission for use must always be obtained from Springer. Permissions for use may be obtained through RightsLink at the Copyright Clearance Center. Violations are liable to prosecution under the respective Copyright Law.

The use of general descriptive names, registered names, trademarks, service marks, etc. in this publication does not imply , even in the absence of a specific statement, that such names are exempt from the relevant protective laws and regulations and therefore free for general use.

While the advice and information in this book are believed to be true and accurate at the date of publication, neither the authors nor the editors nor the publisher can accept any legal responsibility for any errors or omissions that may be made. The publisher makes no warranty, express or implied, with respect to the material contained herein.

Cover design: Beatrice &, Milan

Typesetting with LaTeX: PTP-Berlin, Protago TeX-Production GmbH, Germany (www.ptp-berlin.eu)

Springer-Verlag Italia S.r.l., Via Decembrio 28, I-20137 Milano
Springer is a part of Springer Science+Business Media (www.springer.com)

SEP 0 8 2014

LIBRARY
UWO-FOX CITIES

This book is dedicated to all men and women who believe in a better future and make the daily difference

TP
359
.H8
Z56
2011

Foreword

Renewable energies will play a very significant role in our energy future. This is why I lead the Laboratory of Solar Systems at INES in France and why Gabriele chose to work on photovoltaic systems in the same team. With the decreasing prices of photovoltaic modules and systems, the grid parity has already been reached in some regions of Southern Europe, which means that solar electricity is already able to compete with conventional electricity in terms of selling price. Within the next ten years, solar photovoltaic energy will even be able to compete with conventional electricity in many regions. A similar picture can be drawn for wind energy systems.

However, there is a market barrier coming from a big difference between renewable and conventional energy sources: Solar systems only produce energy when the sun is shining. Wind energy varies with the wind speed. Traditional electricity operators, especially in France, therefore tend to call renewable energies as "fatal" energy sources, because they do not have the tools to control them.

In times when the market penetration of renewable energy is rather low, these fluctuations are not relevant. However, once this penetration to the grid becomes higher, innovative solutions are needed to assure a reliable grid service to the customers. This is very important for the energy supply on an island and also crucial for continental grids with high renewable energy penetration, as we can see today in Germany, for instance.

A first solution might be the massive matching of the electricity demand with the profile of solar energy generation. However, this cannot be done for the complete electricity demand. And matching the demand to fast fluctuations is even more difficult. This is why we have to prepare a second solution – the integration of energy storage. Hydrogen is one promising storage option, as it can be used for both storage and transportation of energy. This is what the book is exploring and what the authors have been researching on for years. I am certain that the reader can find here an interesting introduction on renewable energy systems with hydrogen and how hydrogen can be an interesting vehicle to increase the market penetration of renewable energies.

Le Bourget du Lac, November 2011

Jens Merten
Head of Laboratory for Solar Systems
Institut National de l'Energie Solaire (INES)

Preface

It is just a matter of time before fossil fuels become completely depleted or too uneconomical to retrieve. In light of this development, the current fossil fuel era is meant to draw to an end. If on top of this problem of diminishing availability we also add the environmental pollutions the fuels have caused, it is understandable why we must soon find ways to end the current period and enter a new energy era.

Hydrogen is regarded as one of the most promising candidates capable of assuming a leading role during this historical transition. Needless to say, the energy needed to obtain hydrogen cannot be provided by fossil fuels. It is therefore necessary to turn to renewable energy sources which are inexhaustible and cause as little environmental impact as possible. Amongst these sources, the authors consider solar energy to be one of the best choices for reasons that will be elaborated in the following of the book.

The work is structured into eleven chapters to present the readers with advanced knowledge on the functioning and the implementation of a solar hydrogen energy system, which combines different technologies efficiently and harmoniously to convert renewable energies into chemical energy stored in the form of hydrogen and then to a much more exploitable form of energy, electricity.

Chapter 1 introduces the macro-economical, technical and historical aspects of the new hydrogen-based energy system. Chapter 2 describes the physical and chemical properties of hydrogen, its production, application, the degenerative phenomena and the compatibility of the materials employed to handle hydrogen storage and transportation. Chapter 3 explores in detail the behaviour and the modelling of electrolysers and fuel cells. Chapters 4 and 5 describe the technical foundations of photovoltaic and wind energies. Chapter 6 discusses other potential renewable energy sources for hydrogen production. Chapter 7 addresses another important issue of the whole process: the storage of hydrogen. Chapter 8 provides more information on the chemical storage in standard batteries and other more advanced alternatives. Chapter 9 finally examines in detail the actual complete implementation of the hydrogen system and simulates the system behaviour with the help of mathematical models. Chapter 10 proceeds to present some of the most interesting real-life applications, while Chapter 11 draws the final conclusions. At the end of every chapter are listed the relevant references for readers who wish to further explore the topics.

This book has been conceived with the goal to share the science and technology of solar hydrogen energy systems and to help building a new sustainable energy economy. We hope that we will succeed.

We are grateful to Simone Pedrazzi for helping develope the models and the simulations in parts of the book; and to Andrea Zanni, Secretary of the Board of Wikimedia Italy, for verifying the correct use of the Creative Commons licence of the images taken from the Wikimedia database.

The authors are also indebted to Pei-Shu Wu whose translation and editing have greatly improved the final draft of the book.

Finally, we would like to thank Francesca Bonadei, Maria Cristina Acocella and Pierpaolo Riva from Springer Italia, for their support during the final stages of the publication.

Bologna, September 2011

Gabriele Zini
Paolo Tartarini

Contents

Acronyms

AC	Alternate Current, Activated Carbon
AE	Alkaline Electrolyser
AFC	Alkaline Fuel Cell
BET	Brunauer-Emmett-Teller
BoS	Balance of System
CAES	Compressed Air Energy Storage
CHP	Combined Heat and Power
COP	Coefficient of Performance
DC	Direct Current
DL	Double Layer
DOE	Department of Energy
EDL	Electrical Double Layer
EL	Electrolyser
FC	Fuel Cell
FF	Filling Factor
GHG	Greenhouse Gas
HA	Hydrogen Attack
HC	Hydrocarbon
HCV	Higher Calorific Value
HE	Hydrogen Embrittlement
HFL	Higher Flammability Limit
HHV	Higher Heating Value
HTE	High Temperature Electrolysis
HTS	High Temperature Shift
IEA	International Energy Agency
IEC	International Electrotechnical Commission
LCV	Lower Calorific Value
L-F	Langmuir-Freundlich (equation)
LFL	Lower Flammability Limit
LHV	Lower Heating Value
LIB	Lithium-Ion Battery

LTS	Low Temperature Shift
MCFC	Molten Carbonate Fuel Cell
MCP	Measure, Correlate, Predict
MPPT	Maximum Power Point Tracking
MWCNT	Multi-Wall Carbon Nano-tube
NBP	Normal Boiling Point
OTEC	Ocean Thermal Energy Conversion
PAFC	Phosphoric Acid Fuel Cell
PDF	Probability Distribution Function
PEM	Proton Exchange Membrane, Polymer Electrolyte Membrane
PEMFC	Proton Exchange Membrane Fuel Cell, Polymeric Electrolyte Membrane Fuel Cell
PLC	Programmable Logic Controller
PM	Particulate Matter
PME	Polymeric Membrane Electrolyser
PV	Photovoltaic
QoS	Quality of Service
RES	Renewable Energy Source
SHC	Specific Heat Capacity
SHE	Standard Hydrogen Electrode
SHES	Solar Hydrogen Energy System
SMES	Superconducting Magnetic Energy Storage
SMR	SteaM Reforming
STP	Standard Temperature and Pressure
SOC	State Of Charge
SOFC	Solid Oxide Fuel Cell
SPE	Solid Polymer Electrolyser
SRC	Specific Rated Capacity
SWCNT	Single-Wall Carbon Nano-Tube
TM	Trademark
TSR	Tip-Speed Ratio
UC	Ultra-Capacitor
UPS	Uninterruptible Power Supply
USD	United States Dollar
VRB	Vanadium Redox Battery
VRLA	Valve Regulated Lead-Acid

1

Introduction

The decreasing availability and the negative externalities of the fossil fuels have posed a prominent risk to our ecosystem. Hydrogen can replace these traditional fuels as one of the most promising energy carriers for the future energy economy. This chapter discusses the sustainability of energy sources and demonstrates how a new energy system based on hydrogen and renewable sources can be technically and economically feasible.

1.1 The Current Situation

Nearly 88% of the current energy economy relies on fossil fuels which are not only diminishing rapidly in quantity but also damaging the ecosystem significantly. It is necessary to adopt a fresh mindset to find solutions to the problems and to devise a future with a more secure and sustainable energy supply. To achieve this requires a different energy system based on natural renewable energy sources or safe and clean nuclear technologies.

Since it takes hundreds of millions of years to generate fossil fuels, it is impossible to expect that, at the current consumption rate, such resources will replenish themselves rapidly enough for utilization. Such energy sources therefore cannot be considered renewable as they cannot regenerate in a reasonable time frame. On the contrary, the sources that are defined as *renewable energies* come from a natural process that constantly repeats itself over a short period of time. Among many of these renewable sources, for example, is the electromagnetic energy from the Sun that reaches our planet every day. Other examples include the gravitational forces between the Moon and the Earth, and the geothermal energy inside our planet.

Energy can also be provided by nuclear technology, particularly through fusion power plants that try to recreate on Earth the process that takes place inside the stars. This however still poses formidable technological challenges that will probably not be solved in time before the final depletion of fossil fuel, let alone the damages the fuels continue to bring to the environment in the interim. Meanwhile, the

Zini G., Tartarini P.: Solar Hydrogen Energy Systems. Science and Technology for the Hydrogen Economy. DOI 10.1007/978-88-470-1998-0_1, © Springer-Verlag Italia 2012

current nuclear fusion technology still presents many disadvantages and safety risks that many consider to outweigh its benefits. For these reasons, at the moment the main attention is focusing on better developing and exploiting the renewable energy sources. There are still vast technical and economic challenges to overcome before a new energy regime is established to fully replace the current fossil fuel-based economy. Besides, this transfer will also bring about major institutional changes as well as a complete and utter paradigm shift in our life style and in the international power equilibrium in the next few decades.

1.2 The Peak Oil Theory

Since fossil fuels are bound to be exhausted, it is essential to have a clear understanding on the current extraction and consumption pattern of this resource before discussing new energy sources that can stand for replacement.

In the 1950s, the American geologist M. K. Hubbert developed a theory named *Peak Oil*, which states that the extraction pattern of the petroleum and other combustibles follows a bell-shaped curve. The trend of the curve shows that the quantity of the discovered and extracted oil increases over the years and reaches a maximum amount, before declining gradually with a symmetrically mirrored trajectory. The concept behind this model is that the availability of fossil fuels is limited, either due to the decreasing new oil reserve discoveries or to the increasing costs of extracting remaining few oil in the existing fields.

The curve is described by the *logistic growth model* equation as:

$$Q(t) = \frac{Q_{max}}{1 + a \exp(bt)} \tag{1.1}$$

where Q_{max} represents the total amount of available sources, $Q(t)$ is the amount of the production accumulated so far and a and b are the constants obtained from the model of crude oil production decrease in the USA from 1911 to 1961.

Many different research data all indicate that Hubbert's model indeed matches closely with the actual production pattern in many oil producing countries over the years. In Figure 1.1, the model is superimposed to the actual recorded oil production trend in the USA between 1910 and 2005 and it is evident how these two curves follow closely one after the other. Other countries have also demonstrated the same production tendency over the years. For example, Indonesia as an OPEC member country has shifted from being an oil-exporting country to an importing one, with a production curve equally similar to Hubbert's model.

At the moment, the forecast based on a static consumption[1] of oil predicts a life cycle of another 50 years for this fuel. This estimate, however, just like others, suffers from a great deal of inaccuracy due to the fact that the actual quantity of the

---

[1] *Static consumption* is intended as a fixed consumption remaining at the current level, independent from the variable world consumptions which are expected to have a net increase and exclude the possibility of discovering new exploitable reserves.

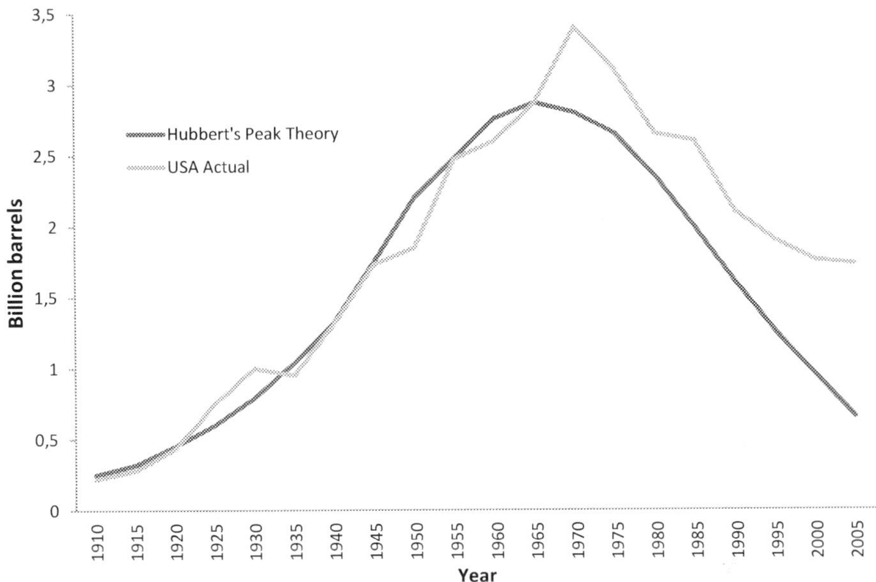

Fig. 1.1. The Hubbert's model and the USA oil production curves

oil reserves is protected in fear of causing global financial panics. Even many statistics provided by international organizations tend to be incorrect for the lack of complete data.

There are also other concerns to consider regarding the efforts invested in discovering new oil reserves. The major oil companies in the world have stated that the ever-increasing rate of consumption will not be sustained even by the finding of new sources. Such reserves are not only difficult to locate, but also require much higher extraction investment, such as building longer pipelines across politically unstable countries with the risks of undergoing terrorist attacks, or sustaining higher refining costs if the reserves contain oil of lower quality. The continuous speculation of the world market on the prize of the crude oil does not help stabilising the situation either. Furthermore, when the oil production begins to decrease according to the decline in the Hubbert's curve, the marginal costs of extraction will start to rise and render the production highly undesirable even before the last drop of oil is exploited. This will force the oil companies to run the risks of less certain investment returns and they will shift the burden to the final users with higher market prices.

The approximation of the Hubbert's curve is also valid for other types of combustibles apart from petroleum. For example, if the life cycle of methane gas according to the current consumption is estimated to be around 65 years, for carbon fossil it is longer than 200 years before its exhaustion. Although it is true that there is no immediate danger for the traditional combustible sources to run out, it is important to take into consideration that most of the fossil fuel reserves are located in politically and socially unstable countries, which results in a situation very similar to an oligopoly with the problems inherent in this type of market. This means that severe supply shortage is more likely caused by political turbulences than by actual fuel unavailability.

Apart from the concerns of reserve availability, an even more pressing problem is the destruction on the global ecosystem caused by the thermodynamics of carbon-based fuel combustion. This will be discussed in the next section.

1.3 Forms of Energy Sources and Environmental Impact

The Sun provides energy to the Earth in the form of electromagnetic radiation. Such energy interacts with the Earth's ecosystem and is transformed into different forms, such as biochemical energy accumulated in the organic systems and the potential energy stored in the movements of air or water masses. Other types of energy are available from the planet itself or from the gravitational interaction with the Moon (see Table 1.1).

Table 1.1. Origins of types of renewable energy available on Earth

Origin	Energy	Type
Sun	Electromagnetic	Radiation (thermal and photovoltaic)
		Potential (water cycles)
		Potential (wind, waves)
		Biochemical (biomasses)
Earth	Radioactivity	Thermal
Moon	Gravitational	Potential (tides)

From a thermodynamic point of view, the sources that produce minimum entropy per energy unit are gravitational forces, nuclear fusion and solar radiation. These energies are manifested on Earth in the forms of electromagnetic energy, wind and tidal movements, oceanic thermal exchanges, marine currents, water cycles, geothermal sources, biomass and nuclear fusion. All of them generate limited environmental impact and possess considerable potential which has not yet been fully exploited on a large scale.

In the meantime, up until now the world has obtained most of its energy from the combustion of fossil fuels. Such combustion produces by-products which cause severe pollutions of air, soil and water sources. It also emits billion tons of CO_2 into the air per year, together with other harmful substances like nitrogen and sulphur oxides. CO_2, when freed in the atmosphere, prevents the heat accumulated on the Earth's surface from being released into the space and creates a harmful greenhouse effect, among other externalities which will be explored below.

Many scientific and industrial organisations have studied the phenomenon of average temperature increase in correlation to the augmented concentration of *green-*

house gases (GHG) in the atmosphere. The *Intergovernmental Panel on Climate Change* has predicted that for the next 100 years the average temperature on the planet will rise by between 1.4% and 5.8%, specifying that the emission of climate-altering gas is the most probable cause for such increase. Other national organizations, like the *German Advisory Council on Global Change*, have also foreseen changes on a global scale across different climate zones after a series of extreme meteorological phenomena, caused by our treating the Earth and the atmosphere as a dumping ground for climate-altering gases.

However, many scholars have voiced a different opinion over whether the temperature rise is caused by human-related CO_2 venting in the atmosphere [30]. The difficulties in correctly calculating the CO_2 concentration in the atmosphere and in evaluating the impact of solar cycles on temperature variations are some of the reasons behind these doubts. In any case, the scientific community is pre-empting the worst-case scenario of the global warming and is suggesting governments to adopt an appropriate strategy to confront the situation. Such strategy is also beneficial since our society will be forced to find substitutes to the fast-diminishing fossil fuels that are still causing damages to human health and the environment.

One solution that prevents the CO_2 from being accumulated in the atmosphere is to capture the released gas with the sequestration technology. However, there is still much room for improvements for this difficult and expensive procedure. First of all, the current emission of CO_2 has been confirmed to be around 6 Gt per year, while to stabilize the current climate condition this volume needs to be reduced to 2 Gt per year. The sequestration technology will need to become operationally mature enough to process such amount each year. Furthermore, on the economic side, casting aside the possibility of generating revenues, this technology entails rather high costs that no business would be willing to sustain if not supported by some governmental incentive policies. In order to find a more attractive and revenue-generating solution to the problem of CO_2 sequestration, it has been suggested to use carbon in a more productive way: instead of storing the carbon in an ever growing volume on a permanent basis, it can be used as a raw material for construction purposes, so that each energy production plant can be regarded as an open-air coal mine of its own. Moreover, according to W. Halloran [10], such approach also allows the released carbon to obtain an economic value capable of generating profits and further encourages the development of CO_2 sequestration technology for sales and trading purposes. The revenues thus generated could also be regarded as part of the avoided costs of environmental pollution, further improving the overall economic benefits of the approach. On another note, since the procedure of CO_2 sequestration increases the cost of fossil fuel energy production roughly by a few cents of USD per kWh, other types of carbon-free energy will become more attractive and more cost-competitive in comparison.

Apart from CO_2, fossil fuel burning also contributes to more than half of the global emission of sulphur oxide, nitrogen oxide and other heavy metallic elements and particulate matters. These substances are deemed to be responsible for causing a wide range of diseases. In 1997, M.A.K. Lodhi [16] estimated that the externalities costs of environmental damages provoked by the combustion of carbon were around 990 billion dollars, while the costs from the consumption of oil and and natural gas are 950 and 400 billion dollars respectively. It appears though that those numbers were underestimated, since at that time the complete extent of environmental and human health damages had still not been fully analysed and understood.

The ecosystem in which we are living now possesses a mechanism that for thousands of years has processed and stored an enormous quantity of CO_2 and other harmful substances underground or in the oceans. It appears that this mechanism of purification was disrupted abruptly by the outbreak of the First Industrial Revolution at the end of the Eighteenth century. If the present rate of fuel consumption continues, in a few years the mankind will have exhausted the sources that the planet had accumulated during the course of 400 to 500 million years. The atmosphere will be filled again with greenhouse gases, heavy metallic elements, sulphur and other particles that had been previously sequestered. The risk is eminent. A sustainable energy system is the only solution to effectively reduce the externalities of fossil fuel combustion and to restore the ecosystem to its former balance.

1.4 Sustainability of an Energy System

According to P. Moriarty and D. Honnery [22][23][24], an energy system is sustainable if it complies with the following criteria:

- the *Energy Ratio*, calculated as the energy source output divided by the energy source input, is higher than one;
- the marginal increase in the energy source input produces proportional or more than proportional marginal increase in the energy output;
- no negative externalities are incurred by the use of such energy source, or such externalities are more than sufficiently compensated.

The first criterion indicates that the quantity of the energy generated during the process is larger than the amount of energy invested. The second suggests that, when the input is augmented, the output is increased as well at least proportionally. Finally, this energy system should be socially acceptable without causing any negative environmental impact, such as air, soil and water pollutions, deforestation, the increase of the amount of acids in the ocean, etc. Any energy system that does not meet these criteria cannot be considered sustainable, will be highly inefficient and even cause damages to the environment and mankind.

Many renewable energy sources, fortunately, are able to satisfy these prerequisites. However, they are also characterised by their volatile performance, meaning that the amount of energy yielded can vary significantly within a very short period of time. For instance, the intensity of solar radiation reaching the ground can change within a day, during a week or a month, depending on the time of the day, the solar elevation angle and the meteorological conditions. The wind and other renewable sources unfortunately share the same volatility. Therefore, an economy intended to relay on this type of sources should tackle this daily and seasonal unevenness and smooth the fluctuations of the energy supply in order to provide a more reliable and programmable energy production structure. The solution lies in efficiently storing the energy produced and managing its distribution to the grid and to the final user. There are many different methods of energy storage available and as will be demonstrated in the following chapters, hydrogen is a highly-advantageous and competitive option.

1.5 A Hydrogen New Energy System

Hydrogen is a widespread element and commonly found in water in its oxidised form. By inputting an external energy source, water can be decomposed into its two main components, oxygen and hydrogen. By spontaneous recombination of hydrogen and oxygen, energy and water are obtained and the latter is ready to be re-used to start the cycle again. Separation of water into hydrogen and oxygen by means of electricity is called *electrolysis*; the *electrolyser* is the device to perform such task. Combustion of hydrogen and oxygen to obtain water and energy is performed in the reverse direction of the reaction of electrolysis, in a device named *fuel cell*. While energy needs to be supplied in order to perform electrolysis, the recombination of hydrogen and oxygen can provide energy instead.

Hydrogen, just like electricity, is an *energy carrier* capable of storing energy converted from the primary sources such as natural gas, oil, coal, nuclear reaction and other renewable energy sources. It can be used directly to produce electricity and heat and has the ability to replace hydrocarbon fuels in a wide variety of applications. Also, compared to traditional fuels, hydrogen contains a higher energy density per weight unit; however it has a very low energy density per volume unit which poses technological challenges that still need to be solved and will be discussed in the course of the book.

There are three types of energy sources applicable to hydrogen production:

- fossil fuels;
- nuclear energy;
- renewable energies, i.e. hydroelectric and geothermal energies, biomass, wind energy, photovoltaic and solar thermal energies.

As previously mentioned, the production of hydrogen must avoid using fossil fuels or traditional nuclear energy technology in order to fulfil the third Moriarty-Honnery criterion discussed in Section 1.4. Therefore, the remaining third source is the only option with the potential to construct a truly sustainable energy system.

1.6 Scenarios for the Future

A *hydrogen-based economy* is defined as a world-wide energy system in which the most commonly used energy carrier is hydrogen. A successful energy carrier must be able to be employed in both *stationary* and *non-stationary* uses. Stationary systems refer to static applications of energy in fixed locations or on-site operations, while non-stationary systems are generally related to transport of people and goods.

An example of a stationary hydrogen energy system is the project called *RenewIslands* [6]. It is a research programme initiated on several European islands in order to develop and verify the different innovative methods to produce and store hydrogen. One of the test benches was installed on the isle of Madira, Portugal, which comprises a wind farm of 1.1 MW, an electrolyser of 75 kW, a hydrogen storage system of 300 kWh and a fuel cell of 25 kW. The encouraging results of the project demonstrate that considering the entire life cycle of the hydrogen energy system, it is indeed more economically convenient to store energy in the form of hydrogen than in traditional batteries.

Non-stationary applications of hydrogen need to be fully developed as well in order to replace fossil fuels entirely in this aspect. Only then can the transition from the current era to the new hydrogen economy be considered complete. The key is to construct a widespread hydrogen supply network for automotive uses. This can be achieved in a way similar to how the fossil fuel distribution network was born in the beginning of the 20th century. In 1908, when Henry Ford started the mass production of Model T, there was no extensive fuel distribution system yet mainly due to the lack of customers. Later on, as the demand started to grow and the market expanded, the suppliers had to increase their production capacity and the retailers had to come up with a complementary fuel-distributing solution, from tapping gasoline into the empty containers brought by the customers to positioning horse-drawn carriages loaded with gas pumps in the busiest streets. It was not until the 1920s did the gas station service as we know today start to grow and spread. These gas stations had limited facilities and ran on low management budget, but they could add other services to provide for the passing travellers' needs in order to create differentiation and build up a profit-generating portfolio. Just as how this fuel distribution infrastructure evolved in a gradual manner, so will the development of the future hydrogen supply and management network, which furthermore can take advantage of an already consolidated logistic structure able to adapt fairly quickly to the characteristics of this new fuel.

Another feature of the future hydrogen economy is its economic accessibility. At the moment, the commercial exploitation of fossil fuels is possible only for large businesses that have the economic capacity to sustain the costs of extraction, production and distribution. Hydrogen production, in contrast, is considerably easier and much more affordable; even home-made hydrogen produced according to personal and individual household needs can no longer be a dream. This freedom from the oligopoly of energy sources will bring profound changes to our social contexts and generate benefits that we can only imagine today.

Researches in different countries are already under way to study the impacts of hydrogen economy on the respective local context. According to Abdallah [1] from the University of Alexandria, Egypt, using solar energy as the renewable source to produce hydrogen would allow Egypt to become an energy exporting country, with the new exported fuel being hydrogen instead of oil. In Europe, Danish scholars Lund and Mathiesen [17][18] have also undertaken a study to evaluate the feasibility and the eventual strategy to reform the Danish energy landscape with renewable sources. Denmark is an oil-exporting country. Since in a few decades the oil reserves in the country will run out, Danish energy experts have begun to explore how to transform their energy system actively and intelligently, with the multiple goals of maintaining energy supply autonomy, reducing environmental impact and at the same time improving industrial development. The three fundamental technological aspects examined in the research are: to improve energy efficiency by regulating the consumption more effectively (on the demand side), to increase energy production efficiency (on the supply side), and to completely replace fossil fuels with renewable sources (again on the supply side). The two most difficult challenges are identified as the integration of the intermittent renewable energy sources into the grid and the supply of energy to non-stationary applications. All in all, the research results have confirmed that a transition towards an energy system based completely on renewable sources is possible. Gradually, the Danish energy economy will go through an intermediate phase around the year 2030 when 50% of the energy sources will be renewable and eventually, the 100% fossil-to-renewable replacement will be achieved by the year 2050, with a total CO_2 emission reduction amounting to as high as 80%.

1.7 Alternatives to Hydrogen

There are also other alternatives that can supplement or replace an energy system based exclusively on hydrogen. As in the case of Denmark, for example, it is feasible to adopt a mixed system (i.e. hydrogen and biomass) or other methods in order to take as much advantage as possible of the combined technologies and to optimize the equilibrium between energy supply and demand. Another possibility is the conversion of electricity to heat through combined heat and power (CHP) plants, heat pumps or electric boilers in order to create an integrated and flexible energy network with a higher overall efficiency. These systems can all work side by side with hydrogen or even become an alternative to an entirely hydrogen-based system.

Another alternative worth considering is silicon, an abundant and commonly-found element on our planet. It can be obtained via economically efficient procedures that do not release carbon-based by-products. It can also be used in exothermic processes with oxygen and nitrogen and can be transported very safely. Furthermore, silicon-based components are easy to recycle and can also be used to produce hydrogen with simple reactions with water or alcohol, apart from acting as an intermediary products in many applications.

Aluminium can also be used as an energy carrier. After removing the outer oxide layer, the pure aluminium can react with water in a combustion process that generates alumina, hydrogen and sufficient heat which can be applied in CHP cycles. Additionally, the hydrogen yielded from this combustion can be used in a fuel cell to provide even more electricity and/or heat. In this regard, aluminium bars can become a potentially viable energy storage device to be used for instance in aerospace technology.

References

1. Abdallah M A H, Asfour S S, Veziroğolu T N (1999) Solar-hydrogen energy system for Egypt. Int. J. Hydrogen Energy 24:505–517
2. Auner N, Holl S (2006) Silicon as energy carrier - Facts and perspectives. Energy 31:1395–1402
3. Anthony R N, Govindarajan V (2003) Management Control Systems. McGraw-Hill, Boston
4. Brealey R A, Myers S C (2003) Capital Investment and Valuation. McGraw-Hill, New York
5. Cavallo A J (2004) Hubbert's petroleum production model: an evaluation and implications for world oil production forecasts. Natural Resources Research, International Association for Mathematical Geology 4 (13):211–221
6. Chen F, Duic N, Alves L M, Carvalho M da G (2007) Renewislands – Renewable energy solutions for islands. Renewable & Sustainable Energy Reviews 11:1888–1902
7. Covey S R (1989) The 7 Habits of Highly Effective People. Simon & Schuster Inc., New York
8. Franzoni F, Milani M, Montorsi L, Golovitchev V (2010) Combined hydrogen production and power generation from aluminum combustion with water: Analysis of the concept. Int. J. Hydrogen Energy 35:1548–1559
9. Hafele W (1981) Energy in a finite world: a global systems analysis. Ballinger, Cambridge, MA
10. Halloran J W (2007) Carbon-neutral economy with fossil fuel-base hydrogen energy and carbon materials. Energy Policy 35:4839–4846
11. Harrison G P, Whittington H W (2002) Climate change – a drying up of hydropower investment? Power Econ. 6 (1):651–690
12. Intergovernmental Panel on Climate Change (2001). Climate change 2001: mitigation. Cambridge University Press, Cambridge, UK
13. Jensen T (2000) Renewable Energy on small islands. 2nd ed. Forum for energy and development
14. Lackner K S (2003) A guide to CO2 sequesterization. Science 300:1677–1678
15. Lehner B, Czisch G, Vassolo S (2005) The impact of global change on the hydropower potential of Europe: a model-based analysis. Energy Policy 33:839–855
16. Lodhi M A K (1997) Photovoltaics and hydrogen: future energy options. Energy Convers Manage 18 (38):1881–1893
17. Lund H (2007) Renewable energy strategies for sustainable development. Energy, 6 (32):912–919
18. Lund H, Mathiesen B V (2009) Energy system analysis of 100% renewable energy systems - The case of Denmark in years 2030 and 2050. Energy 34 (5):524–531
19. Meir P, Cox P, Grace J (2006) The influence of terrestrial ecosystems on climate. Trends Ecol. Evol. 56:642–646
20. Melaina M W (2003) Initiating hydrogen infrastructures: preliminary analysis of a sufficient number of initial hydrogen stations in the US. Int. J. Hydrogen Energy 28:743–755
21. Melaina M W (2007) Turn of the century refueling: A review of innovations in early gasoline refueling methods and analogies for hydrogen. Energy Policy 35:4919–4934
22. Moriarty P, Honnery D (2005) Can renewable energy avert global climate change? In Proc. 17th Int. Clean Air & Environment Conf. Hobarth Australia
23. Moriarty P, Honnery D (2007) Intermittent renewable Energy: the only future source of hydrogen? Int. J. Hydrogen Energy 32:1616–1624

24. Moriarty P, Honnery D (2007) Global bioenergy: problems and prospects. Int. J. Global Energy Issues 2 (27):231–249
25. Nilhous G C (2005) An order-of-magnitude estimate of ocean thermal energy resources. Trans. ASME 127:328–333
26. Penner S S (2006) Steps toward the hydrogen economy. Energy 31:33–43
27. Rifkin J (2002) The Hydrogen Economy. Penguin, New York
28. Romm J (ed.) (2004) The Hype about Hydrogen: Fact and Fiction in the Race to Save the Climate. Island Press, Washington
29. Scheer H (1999) Energieautonomie – Eine Neue Politik für Erneuerbare Energien. Verlag Antje Kunstmann GmbH, München
30. U. S. Senate Minority Report: More Than 700 International Scientists Dissent Over Man-Made Global Warming Claims Scientists Continue to Debunk Consensus in 2008 & 2009 (2009). U.S. Senate Environment and Public Works Committee Minority Staff Report. Original Release: December 11, 2008. Presented at the United Nations Climate Change Conference in Poznan, Poland. Update March 16, 2009
31. Veziroğlu T N (2008) 21st Century's energy: Hydrogen energy system. Energy Conversion and Management 7 (49):1820–1831

2

Hydrogen

The physical properties of hydrogen make it an important potential candidate to substitute fossil fuels in the next energy economy. Hydrogen can be produced via traditional and novel technologies applied in chemical and electrochemical processes. However, being the smallest molecule of all, hydrogen poses serious challenges to the components of its storage and transport systems. For this reason the materials of the pipes and the valves must be carefully selected, while the joints and the connections need to be sealed properly to avoid problems of degradation and leakage.

2.1 Hydrogen as Energy Carrier

The idea to use hydrogen as fuel is very old. It dates back to when hydrogen was first identified and isolated as a chemical element by Cavendish in 1766 and received its name (meaning *water-former*) from Lavoisier in 1781. Jules Verne even described hydrogen as the coal of the future in his fiction *The Mysterious Island*. In 1820 Cecil built a hydrogen generator and in 1923 Haldane managed to produce hydrogen with a windmill. In 1839 Grove conducted the earliest experiments on the use of hydrogen in a prototypal fuel cell, while in 1870 Otto used a mixture containing 50% of hydrogen in his experiments for the first internal combustion engine. As for the applications in the means of transportation, the element was used in air balloons and airships because of its lower density compared to air but was subsequently replaced by helium due to its high flammability. In 1938, Sikorski successfully used it to power the propeller for helicopters.

Nowadays, hydrogen is used as a fuel only in aerospace applications, where liquid O_2 and H_2 are combined together to yield the massive amount of energy required for the spaceships to exit the atmosphere, as well as the electricity needed for the entire crew and instrumentation. Even though the ability of hydrogen to provide energy has been understood for a very long time, in 2001 the quantity of hydrogen applied for such purpose only amounted to 2% of the fossil fuels consumed, with an estimated cost per GJ three times as much as that of the natural gas.

Zini G., Tartarini P.: Solar Hydrogen Energy Systems. Science and Technology for the Hydrogen Economy. DOI 10.1007/978-88-470-1998-0_2, © Springer-Verlag Italia 2012

Although hydrogen has a very high energy density per unit weight, it is considered as an *energy carrier* or a secondary energy source, as opposed to a primary source like wood, petroleum and coal which are immediately available for energy uses. In fact, before its utilization, hydrogen must be obtained in its molecular form (H_2) by chemically transforming other compounds containing the element. Hence, hydrogen is a vehicle able to store energy converted from the primary sources in a chemical form. As an energy carrier, hydrogen can be used in direct combustion processes or, even better, in fuel cells that do not use thermal cycles and therefore are free from the performance limitation of the second law of thermodynamics. By operating on electro-chemical reactions instead, these fuel cells are capable of producing the maximum amount of energy possible thanks to its very high conversion efficiency. Also, the reaction of hydrogen with oxygen releases energy and water without releasing CO_2, a typical by-product of fossil fuel burning. Therefore, in terms of environmental impacts, hydrogen can be the best alternative to fossil fuels, as it drastically reduces the release of climate-changing gases and compounds (such as oxidized nitrogen, sulphur and fine particles) harmful to human health.

Against the above advantages, however, there are still several drawbacks limiting the use of hydrogen as an energy carrier: as previously mentioned, hydrogen needs to be obtained via an extra chemical procedure; furthermore there is still no economic system mature enough to operate on hydrogen on a large scale, even if hydrogen has been extensively employed in the chemical industry for decades.

2.2 Properties

Hydrogen (symbol H, atomic number 1 and electron configuration $1s^1$) is the first chemical element in the periodic table. It is a nonmetal element belonging to group I A with standard atomic weight of 1.00794. At standard temperature and pressure it is a colourless, odourless and tasteless diatomic gas with molecular form H_2.

A hydrogen atom is composed of a proton and an electron. It has two oxidation states ($+1$, -1) with an electronegativity of 2.2 on the Pauling scale. Its highly reactive sole electron orbiting around the nucleus is capable of building covalent bonds with other hydrogen atoms to reach the stable configuration of a diatomic molecule. The covalent radius is around 31 ± 5 pm and the Van der Waals radius is 120 pm. The ground state energy level is -13.6 eV and the ionization energy is 1312 kJ/mol. There are two hydrogen isotopes: *Deuterium*, with 1 proton and 1 neutron and *Tritium* with 1 proton and 2 neutrons.

Every proton in a hydrogen molecule has its own type of spin which results in the existence of two types of H_2 molecules:

- *ortho-hydrogen*: the protons of the two atoms have the same spin;
- *para-hydrogen*: the protons of the two atoms have an opposite spin.

Hydrogen is the most commonly-found element in the universe and it is also one of the main components of stars and interstellar gases. For example, an astronomical analysis conducted on the light transmitted by our closest star, the Sun, shows

Table 2.1. Physical properties of hydrogen

Molecule	H_2
Phase at STP	Gas
Melting point	14.025 K
Boiling point	20.268 K
Molar volume	$11.42 \cdot 10^-3$ m^3 mol
Enthalpy of vaporization	0.44936 kJ/mol
Enthalpy of fusion	0.05868 kJ/mol
Density	0.0899 kg/m^3
Speed of sound	1270 m/s at 298.15 K
Electronegativity	2.2 (Pauling scale)
Specific heat capacity	14304 J/(kgK)
electric conductivity	N/A
Thermal conductivity	0.1815 W/(mK)
Ionization energies	1312.06 kJ/mol
Lower calorific value	110.9–10.1 (MJ/kg–MJ/Nm^3)
Minimum ignition energy	0.02 mJ
Stoichiometric flame speed	2.37 m/s
Density	0.084 kg/Nm^3
Boiling point	20.4 K
Critical point	32.9 K
Specific heat	14.9 kJ/(kg K)
Flammability limits (by volume percentage)	4–75%

that around 75% of its mass is composed by hydrogen. Hydrogen is also one of most abundant elements on Earth and can be found in a wide range of organic and inorganic molecules like water, hydrocarbons, carbohydrates and aminoacids etc. Pure hydrogen though is very rare on Earth since the gravitational force of our planet is not strong enough to hold on to such light molecules.

Some of the physical properties of hydrogen are summarised in 2.1.

The *hydrogen bond* is a type of weak electrostatic bond that forms when a partially positive hydrogen atom covalently bonded in a molecule is attracted by another partially electronegative atom equally covalently linked to another molecule. It is described in particular as a dipole-dipole interaction when a hydrogen atom shares a covalent bond with highly electronegative elements like nitrogen, oxygen and fluorine which attract valence electrons and acquire a partial negative charge, leaving the hydrogen with a partial positive charge. The bond forms when a relatively strong electropositive hydrogen atom comes into contact with an electron pair of another chemical group or another molecule. For example, in OH there is a partially negative charge from oxigen and a positive charge from hydrogen, therefore OH becomes polarized as a permanent dipole. The energy of the hydrogen bond, which is a few kJ and depends on the local dielectric constant, is weaker than that of the ionic and covalent bonds but generally stronger than or equal to Van der Waals forces. Hydrogen bond is present in both the liquid and the solid states of water and contributes to its relatively high boiling point (as opposed to, say, H_2S which is less polarized).

Another peculiarity about the hydrogen bond is that it keeps the molecules more distant from each other in respect to other types of chemical bonds, which is why ice has a lower density than water. In fact, water molecules float as liquid but form a crystal structure in the solid state as ice. Hydrogen bonds also exists in proteins and nucleic acids and act as one of the main forces to unite the base pairs of the double helical structure of DNA. This bond is fundamental for the equilibrium of our ecosystem. Without it, for example, water would have very different physical properties and the current life forms as we know on this planet would be impossible to exist.

2.3 Production

As previously mentioned, hydrogen molecule H_2 must be produced from other hydrogenated compounds. Some of these procedures have reached industrial maturity while others are still in development:

- *consolidated technologies*: steam reforming of hydrocarbons, gasification of solid fuels and water splitting by electrolysis;
- *alternative methods*: thermochemical water splitting at high temperature, photo-biological reactions, biomass conversions, power-ball and others.

2.3.1 Steam Reforming

Steam reforming (SMR) of hydrocarbons is the most prevalent industrial procedure to produce hydrogen. With the process of transforming methane (CH_4), the endothermic reaction that takes place at high temperature with the use of catalysts is expressed as the *reforming reaction*:

$$CH_4 + H_2O + \text{heat} \rightarrow CO + 3H_2 \tag{2.1}$$

which is followed by the exothermic reaction of *shift*:

$$CO + H_2O \rightarrow H_2 + CO_2 + \text{heat}. \tag{2.2}$$

These two reactions can by synthesized in the combined endothermic reaction:

$$CH_4 + 2H_2O \rightarrow CO_2 + 4H_2. \tag{2.3}$$

Normally part of the heat supply comes from burning the fuel at the start of the reaction and part from consuming the final product. The efficiency of the procedure is defined as the ratio between the chemical energy contained in the produced hydrogen and the energy stored in the supplied methane. This varies from 60% to 85% and the highest performance can be achieved if the consumed heat is recovered. Steam reforming plants are often complex and large structures; in fact they are usually constructed for a production capacity of 105 Nm^3/h. The steam reforming process generates a synthetic gas that contains hydrogen among other pollutants and carbon dioxide. For that reason it is necessary to proceed with a post-treatment to eliminate the pollutants and to separate the carbon dioxide in order to obtain hydrogen molecules with a high degree of pureness.

The SMR process normally follows the phases as described below:

- *Feedstock purification*: removal of sulphur and chloride via molybdenum and cobalt oxide catalysis.
- *Pre-reforming*: initial reforming at low temperature in order to reduce the dimension of the main reformer and to pre-process heavier hydrocarbons. Only CH_4 and CO are present in the final product.
- *Reforming*: the gas and the steam pass through tubes inside a heater with the presence of nickel-based catalysts. The reactions in the heater are endothermic with the heat being supplied by radiation or a burner. This process takes place only at pressure around 30 bar and at a temperature of 850 to 1000 °C.
- *High Temperature Shift* (HTS) conversion: process of converting CO to CO_2 with iron and chromium oxide catalysts at a temperature of 350 °C.
- *Low Temperature Shift* (LTS) conversion: process of converting CO to CO_2 with copper-based catalysts at 200 °C.

2.3.2 Solid Fuel Gasification

This process of *solid fuel gasification* entails the gasification of coal with steam, producing the so-called *water gas*. The complete reaction applied to C is:

$$C + H_2O + heat \rightarrow CO_2 + 2H_2. \tag{2.4}$$

This synthesis gas contains much more pollutants and carbon oxides compared to the SMR procedure performed on methane. Post-treatment plants therefore are necessary but they can be complicated and expensive to build. In terms of environmental impacts, even if gasification allows us to obtain a clean product like hydrogen from the abundant and cheap coal, it does not avoid producing the climate–altering carbon oxide. In fact, considering all the chemical reactions involved from the initial gasification to the subsequent hydrogen combustion, the same quantity of carbon oxide will still be produced as in the simple burning of carbon. This phenomenon in fact exists in all the hydrogen production processes involving hydrocarbons. It is possible to reduce the overall carbon oxide emission only by adopting efficient energy conversion methods which exclude the use of carbon fuels from the beginning, such as fuel cells. The contribution brought forth by the SMR plants and the gasification procedure, however, is that they can render the sequestration of carbon oxide simpler and more feasible.

2.3.3 Partial Oxidation

It is also possible to obtain hydrogen from crude oil residuals and heavy hydrocarbons via a *partial oxidation* reaction like the following:

$$CH_4 + 0.3H_2O + 0.4O_2 + heat \rightarrow 0.9CO + 0.1CO_2 + H_2 \tag{2.5}$$

in which the required heat comes directly from burning part of the fuel with a controlled supply of air at the start of the reaction. For smaller plants, this characteristic together with the almost total absence of catalysts make the whole procedure better than steam reforming, even though it requires the use of pure oxygen and delivers a lower energy conversion performance.

2.3.4 Water Electrolysis

Water electrolysis is the process of separating hydrogen and oxygen from water by the use of electric current, therefore it is a procedure that converts electric energy to chemical energy. Electrolysis of water in particular is very important industrially as it makes it possible to obtain hydrogen and oxygen with a high degree of pureness. The subject will be explored in more depth in the next chapters.

2.3.5 Thermo-Cracking

Hydrogen can also be obtained from hydrocarbons by *thermo-cracking*. The procedure uses a plasma burner at around 1600 °C to separate hydrocarbons into hydrogen and carbon atoms with this reaction:

$$CH_4 \rightarrow C + 2H_2. \tag{2.6}$$

Working with hydrogen molecules and using only electricity and cooling water to stabilize the temperature, pure hydrogen gas can be produced without emitting CO_2. The efficiency of this process is normally around 45%.

In other thermo-cracking processes, the chemical reaction is the same but hydrogen molecules are decomposed at a very high temperature without producing water steam. The heat required (around 8.9 kcal/mol) is supplied by methane combustion but it can also come from using the hydrogen generated in the process, eliminating thus the emission of CO_2. The main technical difficulty lies in finding the suitable catalysts for the reaction, as the traditional ones are easily degraded by carbon residues. The efficiency of this procedure is usually around 70% of the performance of steam reforming on methane.

2.3.6 Ammonia Cracking

Ammonia can be a good hydrogen carrier as it can undergo the cracking process to release hydrogen and nitrogen with the following *ammonia cracking* reaction:

$$2NH_3 \rightarrow N_2 + 3H_2. \tag{2.7}$$

Ammonia is yielded from the chemical reaction between water, methane and steam. Water is then removed along with carbon oxide and other sulphur compounds to obtain a mixture of pure hydrogen and nitrogen that will not corrode the catalysts used in the process. The gas generated from the reaction is then cooled off to obtain liquid ammonia, which is then stored and transported at ambient temperature at 10 atm or cooled off under its boiling temperature of 240 K in non-pressurised containers.

Ammonia is easy to transport and to store, which makes it a very convenient means to move and store hydrogen as well. The only downside is that even the minimum trace of ammonia can cause problems to fuel cells, as it can form carbon compounds which block the electrodes and slow down the fuel cell reaction and performance.

2.3.7 Other Systems: Photochemical, Photobiological, Semiconductors and their Combinations

Hydrogen production is also possible when the sunlight interacts with:

- photochemical systems;
- photobiological systems;
- semiconductors;
- combinations of the above.

In *photochemical systems*, the sunlight is absorbed by molecules in a solution with water often used as the solvent. Since water filters out a significant portion of solar radiation, a *sensitiser* (either a semiconductor or a molecule) is required to absorb the photons carrying sufficient amount of energy so that hydrogen can be produced. The absoprtion of a photon by the sensitiser yields a free electron through the (simplified) reaction that occurs with catalysts:

$$H_2O + Energy \rightarrow H_2 + \frac{1}{2} O_2. \tag{2.8}$$

A serious drawback of such system though is the concurrent production of H_2 and O_2 that can pose potential safety risks when mixed together. The system however has operated on a good efficiency of over 10%

In *photobiological systems*, the light interacts with chloroplasts or algae together with enzymes that are capable of facilitating the production of hydrogen. Such natural systems were among the earliest to appear on our planet. Biochemical cycles, such as photosynthesis and the Krebs cycle in the plants, harvest the energy from solar radiation in order to convert CO_2 into organic compounds, especially into sugars. While the conversion efficiency is merely 2%, the overall leaf surface is large enough for the plants to capture all the energy needed to grow and flourish. A photosynthetic process to produce hydrogen can be developed by modifying the operation conditions of micro-organisms, such as micro-algae and cyanobacteria, or proteins like ferredoxins and cytochromes. Normally, in this process the energy is stored by converting CO_2 into carbohydrates. In anaerobic conditions, the micro organisms synthesise and activate the *hydrogenase* enzyme, which produces H_2 and O_2 when exposed to the light with an efficiency that can be fairly high (i.e. 12%). The reaction entails the intervention of electron donors (D) or acceptors (A) for hydrogenases as in the following:

$$H_2 + A_{ox} \rightarrow 2\,H^+ + A_{red} \tag{2.9}$$

$$2\,H^+ + D_{red} \rightarrow H_2 + D_{ox}. \tag{2.10}$$

This type of system possesses many potential advantages, such as the capacity of auto-organisation and regulation. However, the hydrogen flow generated in this way is limited and the industrial applicability of the process is still to be demonstrated. Real-life applications can be difficult also because the organisms will funcion only in a carefully-defined and optimised micro-climate. The current tendency is to apply genetic engineering techniques to modify the organisms in order to improve their long-term resistance, increase conversion efficiency and permit their applications in normal or even extreme environmental conditions.

Solar hydrogen production can be also obtained by means of *photodegradation* of organic substrates. For example, this reaction shows the process performed on pulluting substances:

$$CH_3\,COOH(aq) + O_2 \rightarrow 2\,CO_2 + 2\,H_2 \tag{2.11}$$

which is exoergonic and hydrogen is produced by oxidation, taking advantage of the decomposition of a potentially harmful substance and generating extra benefits from the process.

In semiconductors, the photons are absorbed by small semiconductor particles suspended in liquids.

The combination of all of the above systems are still under active research and development. For instance, semiconductors are combined together with organic systems in order to improve the response of the material to the photon wavelengths that normally cannot be captured by semiconductors alone.

2.4 Usage

Hydrogen is a very versatile energy carrier. It can release energy in a number of different processes:

- direct combustion;
- catalytic combustion;
- steam production;
- fuel cell operations.

2.4.1 Direct Combustion

A mixture of hydrogen and oxygen with an appropriate proportion plus the presence of a trigger can release thermal energy until one of the two components is fully consumed:

$$H_2 + \frac{1}{2}O_2 \rightarrow H_2O + \text{heat}. \tag{2.12}$$

The burning of hydrogen and oxygen is a traditional method for spacecraft propulsion, whereas the combustion of hydrogen and air is used more frequently in chemical and manufacturing industries. This direct combustion has many advantages. First of all, the wide flammability range of hydrogen allows the combustion of the gas in mixture with other gases and creates a sensible reduction of the maximum flame temperature. Secondly, hydrogen can also replace traditional fuels in volumetric and internal combustion engines. The high speed of hydrogen flame can even benefit internal combustion engines by granting them a very high rotation regime. Furthermore, in comparison with fossil fuel combustion, hydrogen burning releases much fewer, if none at all, pollutants like carbon oxides (CO_x), particulate matter[1], sulphur oxides (SO_x, recognised as carcinogenic agents) and nitrogen oxides (NO_x, irritating but non-toxic agents).

From a thermophysical point of view, hydrogen is often compared to methane. Using methane is certainly a step forward to reducing the production of these pollutants as it does not emit SO_x or particulates and generates a lower quantity of CO, CO_2 and HC thanks to the low C/H ratio. But it still does not reach the zero-emission standard. On the contrary, the lack of sulphur and carbon in hydrogen fuel signifies the total absence of CO, CO_2, HC, particulates and SO_x, with a limited release of NO_x in any case. Table 2.2 shows a comparison of some of the most important physical properties of the two gases in terms of energy production.

The considerable difference between the atomic dimensions of these two gases only accentuates the difficulty to manage hydrogen in the same pipelines and distribution networks built to transport methane. In fact hydrogen has a great capacity of passing through tube junctions, penetrating through the pores and even damaging the material. Existing methane pipings hence cannot be immediately used for hydrogen transportation. Table 2.2 also shows that hydrogen boiling point, critical point and density are all lower than those of methane, complicating the issues of storage, transport and safety management.

The lower heating value of hydrogen is higher than that of methane but lower per unit volume. Hydrogen gas is extremely light and consequently has a reduced energy density per unit volume, while its stored energy per unit mass is very high.

The *flammability limits*[2] of hydrogen range from 4% to 75% while the detonation limits fall between 15% and 59%. The wide span of flammability limits on one hand means that it is possible to burn a poor hydrogen mixture and to produce minimum nitrogen oxide by-products and lower flame temperature; on the other it entails that hydrogen can be ignited even with a very low percentage of air present (as for example when the air infiltrates into the hydrogen carrying pipelines), creating therefore high potential security threats.

---

[1] *Particulate matter* (PM) is a complex mixture of tiny subdivisions of solid matter suspended in a gas or liquid. These particulates are categorised according to their diameters, and those with a diameter smaller then 10μm (PM_{10}) are defined as particles in the *thoracic fraction* capable of entering pulmonary alveoli and causing cancer and permanent lung deficiencies.

[2] *Flammability limits* are defined by the range of volumetric proportions within which a combustible gas mixture is flammable with the presence of an ignition source.

Table 2.2. Comparison among hydrogen, methane and petrol

	Hydrogen	*Methane*	*Petrol*
Mass (g/mol)	2	16	
Volumetric mass (kg/m^3)	0.08	0.7	
Density (kg/Nm^3)	Table 2.1	0.651	
Boiling point (K)	Table 2.1	111.7	
Critical point (K)	Table 2.1	190.6	
Specific heat (kJ/kg K)	Table 2.1	2.26	
Lower heating value (MJ/kg)	110.9	50.7	44.5
Lower heating value (MJ/Nm^3)	10.1	37.8	
Flammability minimal energy, ambient (mJ)	0.02	0.29	0.24
Flammability limits (by volume) (%)	4–75	5.3–15	1.0–7.6
Detonation limits (by volume) (%)	13–65	6.3–13.5	1.1–3.3
Auto-ignition temperature (°C)	585	540	228–501
Flame temperature (°C)	2045	1875	2200
Diffusion coefficient, in air (cm^2/s)	0.61	0.16	0.05
Stoichiometric flame speed (m/s)	2.37	0.43	
Explosive energy (kg TNT/m^3)	2.02	7.03	44.24
Detonation speed, in air (km/s)	2	1.8	

In this sense, hydrogen might seem to be the least favourable kind of fuel for safety concerns, but in reality petrol and diesel have lower flammability levels and just a small quantity of these fuels is sufficient to cause fire. Hydrogen is less flammable than petrol as its auto-ignition temperature is 585 °C as opposed to the 228–501 °C of the petrol. Since hydrogen is the lightest of all chemical elements, it can dilute rapidly into the open space. It is practically impossible for it to self-ignite if not in a confined space. When it burns, hydrogen consumes very rapidly and produces high-pointing flames, which are characterized by their low long-wave thermal radiation. The flame has a very pale colour due to the absence of carbon and consequently does not produce soot. It is almost imperceptible in the daylight if not for its thermal radiation and it is also invisible in the dark.

Because of these properties, hydrogen is capable of venting very quickly[3], as opposed to fuels like petrol, gasoline, LPG and natural gas which are heavier than air and pose a greater danger as they do not vent in the air as fast. An example is that the burning of a vehicle from a petrol leakage lasts from 20 to 30 minutes, while a hydrogen vehicle burns for no more than 1 to 2 minutes. The low thermal radiation of the hydrogen flame means that it can only burn other materials close-by when in direct contact, therefore the combustion time is reduced along with the danger of toxic emissions. Unlike fossil fuels, hydrogen is neither toxic nor corrosive and its potential leak from the fuel tanks does not pollute the soil or underground water sources.

---

[3] From the *Graham's law of effusion*, the gas effusion rate is inversely proportional to the square root of its molecular weight. For instance, hydrogen vents four times faster than oxygen.

The low electric conductivity of hydrogen is the reason why even a very small electrostatic charge is able to ignite sparks and trigger the mixture of hydrogen and oxygen. Contrary to other hydrocarbons which release heat and visible radiation when in combustion, hydrogen flame emits less heat and is practically invisible because of its ultraviolet light emission. This poses a great safety risk on hydrogen gas management as it is difficult to detect the flame or a fire caused by leakage unless a direct contact is made. Leak detection systems hence are extremely important for the safety measures of a hydrogen plant. Currently, to detect the leak, these systems use a thin membrane which changes its optic properties when hydrogen is absorbed, or palladium which alters its electric resistance when in contact with the gas. The detectors developed by a prominent European car manufacturer for its hydrogen vehicle prototypes, for example, automatically open the car windows and the roof when the hydrogen concentration in the air exceeds the safety limit, set at 4%. Other criteria adopted by hydrogen detection systems, apart from sensitivity and precision, are alarm response time, resistance against deterioration and long-term reliability.

2.4.2 Catalytic Combustion

Hydrogen combustion is also possible with the presence of a catalyst, usually with a porous structure, to reduce the reaction temperature. However, compared to the traditional method, *catalytic combustion* requires a greater reaction surface. The only by-product of the reaction is water steam, since no NO_x is yielded thanks to the low temperature. The process is therefore considered clean with very low gas emissions. The reaction speed can be easily controlled by managing the hydrogen flow rate. Since the reaction does not produce flames, catalysed combustion is intrinsically a very safe procedure.

2.4.3 Direct Steam Production from Combustion

The burning of hydrogen and oxygen can bring the flame temperature up to $3000\,^\circ C$ and produce water steam, consequently more water needs to be injected to maintain the desired steam temperature, forming therefore saturated and superheated steam with an efficiency close to 100% without losing any thermal energy. The steam can be used in turbines and industrial and civil applications.

2.4.4 Fuel Cell

The opposite reaction of water electrolysis is the combination of H_2 with O_2 to generate water. This process releases part of the energy of electrolysis that was used to separate water into its elementary components. This happens in a device called *fuel cell*, which will be discussed later in detail.

2.5 Degenerative Phenomena and Material Compatibility

The physical and chemical properties of hydrogen molecule require the use of carefully selected and manufactured materials and components, which renders the construction of the pipelines and other related devices much more complicated compared to other gases. The main reasons are that hydrogen can cause degenerative phenomena on the materials coming into contact, plus the extremely small dimension of hydrogen molecules poses further problems of containment.

2.5.1 Material Degeneration

Certain materials can suffer from damages of degeneration on their mechanical properties when coming into direct contact with hydrogen. This phenomenon is called *Hydrogen Embrittlement* (HE) and the cause is attributed to the penetration of hydrogen molecules in the metal lattice of the material. Hydrogen is highly diffusible and is prone to attack the defects in the material, the so-called *traps*, which usually locate in the structural imperfections such as dislocations, inclusions, grain boundaries, segregations and micro-voids. An entrapped hydrogen atom has a lower potential energy compared to the atoms of the material lattice. The traps are defined as *reversible traps* if the hydrogen contained within does not react with the material lattice and, hence, does not affect the mechanical and physical properties of the structure. Yet more often than never, the entrapped hydrogen will interact with the atoms of the structure or with other hydrogen atoms to form substances capable of causing degeneration. Dislocations in the material can play a determining role as they are able to influence hydrogen movement and to drive it from reversible to irreversible traps. For example, HE can occur in steels subjected to mechanical solicitations, or on metal surfaces with superficial defects caused by plastic deformation. It can also happen when traces of sulphur are present, as sulphur can easily decompose hydrogen molecules into atoms.

The occurrence of HE depends also on other factors, such as temperature, pressure, hydrogen concentration, hydrogen exposure time, the tensile strength and the pureness of the metal, as well as on the micro-structure and the superficial conditions of the material.

The following correlations concerning HE have been observed:

- HE occurs at ambient temperature and can often be neglected when the temperature is above 100 °C;
- there is a positive correlation between the pureness of hydrogen and HE;
- the higher the partial pressure of hydrogen is (usually between 20 to 100 bar), the more probable HE will occur;
- the greater the mechanical stress is, the more likely HE will appear on the points absorbing such stress.

Different from corrosion, hydrogen exposure time for HE is not so significant because the critical concentration to cause the phenomenon can be reached quite quickly.

The deterioration effect from HE can cause hydrogen leaks and further risks of fire, explosion and fractures in the storage structure and in the measuring instruments. Currently, the only way to avoid HE is to choose appropriate materials and to manage the facilities with great care.

When the temperature reaches above 200 °C, many low alloy steels[4] can also suffer from another type of deterioration due to the presence of hydrogen: the phenomenon, known as *hydrogen attack* (HA), refers to the degradation of the material following the chemical reaction between hydrogen and the carbon compounds in the steel. The reaction also yields methane gas as a by-product. This phenomenon can also form micro-cavities that compromise the steel resistance. The damage becomes more serious with the increase of temperature and pressure since this causes hydrogen to diffuse even further into the steel material of the pipelines and relevant components. To prevent this deterioration, it is necessary to use steels containing stabilized carbon compounds, especially those with rich chromium and molybdenum components, to reduce the hydrogen-carbon reactions.

2.5.2 Choice of Materials

The negative effects of HE and HA can be countered by choosing appropriate materials proven to be able to resist such deterioration. Obviously, the choice of material depends on the condition of the application. For example, a hydrogen pipeline extending for kilometres will require materials different from the structures destined, for instance, to store hydrogen in conditions of static pressure and temperature.

Metallic materials like aluminium and its alloy, copper and its alloy (brass, bronze, copper-nickel) and austenitic stainless steel[5] (such as 304, 304L, 308, 316, 321, 347) have proven to display inconsequential deterioration phenomena and therefore are considered by the *European Industrial Gases Association* as suitable for the use of either hydrogen liquid or gas. Other types of steel containing carbon, such as API 5L X52 and ASTM 106 grade B can also be used for the same application. On the contrary, other materials such as iron can suffer from degradation to such an extent that similar utilization becomes impossible.

Materials like Teflon™ and Kel-F™ can be used with gas and liquid hydrogen, while Neoprene™, Dacron™, Mylar™, Nylon™ and Buna-F™ are usable only with gas hydrogen.

---

[4] An *alloyant* is added into the steel to improve its mechanical properties (such as hardness and toughness) and strengthen its resistance to corrosion. The nature and the amount of the alloyants can gradually increase the difficulty of welding. Alloy steels can contain alloyants up to 50% of the total composition for them still to be considered as alloy steels. To better categorize alloy steels with so many different alloyant contents, a division has been drown between low alloy steels, in which no alloyant exceeds 4%–5 % of the total composition, and high alloy steels, in which at least one alloyant has a percentage above such limit.

[5] *Austenitic stainless steel* is a metallic non magnetic allotrope of iron, with alloying element of C, Ni and Cr in different percentages that preserve its *austenitic*, or *face-centered cubic* system structure, with lattice points on the corners and the faces of the cube.

2.6 Components: Pipes, Joints and Valves

Due to the very small dimension of the hydrogen molecule, it is very difficult to construct hydrogen transport pipelines that effectively avoid leaks. It is therefore very important to choose appropriate components.

The pipes which contain superficial defects or constructed with incompatible materials can provide escape routes for hydrogen immediately or over a certain period of time. Such problem becomes more serious as the tube diameter increases, since this also raises the probability of the existence of manufacturing defects like scratches, cracks and impurities that can form weak points for hydrogen to infiltrate and start the embrittlement process.

The pipe thickness also has to be carefully evaluated to ensure its perfect containment function, since thick walls can resist better than thinner ones, especially at junction points. A thin wall often carries a greater risk of being damaged at the joints.

There are three types of methods to connect two pipe ends together: soldering, joining and threading. Soldering guarantees the best sealing effect, but in case the pipes need to be dissembled in the future, conical and cylindrical threading becomes the most suitable solution. However, since threading can manifest containment issues, joining is preferred if soldering is not an option. The joints usually used for hydrogen pipelines are compression fittings (composed by a bolt, a nut, a front and a rear ferrule) that can practically guarantee a perfect sealing.

2.7 Transport

Hydrogen is usually transported as a liquid or a gas. The transport and the storage of the element are two problems strictly correlated to each other, both depending on the final usage and the quantity of gas as well as on the transport distance.

Compressed gas resorts to vessels with protective shells made of HE-resistant materials at a pressure over 200 bar. The transport usually covers a short distance and can be carried out by truck, by train or with short pipelines. Liquefied hydrogen instead is preferably transported in thermally-isolated spherical containers in order to obtain the maximum volume-to-surface contact ratio and to reduce the evaporation rate to less than 1.1%. This method is preferred for long distance travels either by land or by sea in order to amortise the high operation costs. As for the transport in pipelines, although a long distance (around 100 km) can be technically covered, usually the gas ducts are constructed in a shorter length, mostly located within the same site where H_2 is needed.

Considering the criteria such as the quantity of hydrogen and the travel distance to cover, in principle the preferable transport method can be identified as indicated in Table 2.3.

If the travel distance is short and the quantity is low, carrying compressed hydrogen in cylinders can be the most technically and economically convenient choice. As the distance increases, shipping liquid hydrogen by sea can replace the compressed H_2 transport. Vice versa, if the distance remains the same but the quantity increases,

Table 2.3. Choice for H_2 transport methods

| | | Distance | |
		Short	Long
Quantity	Low	Compression	Liquefaction
	High	Compression / Liquefaction	Gas pipelines

using gas pipelines to transport hydrogen gas or liquid is the best choice. If the quantity is high and the distance is long, gas pipelines will be the most convenient method, since the initial costs can be amortized within a reasonable time frame with rather low operational costs. Furthermore, recent studies have shown that for a pipeline longer than 1000 km it can be more economically convenient to carry hydrogen rather than electricity.

In order to evaluate the dangers of hydrogen transport, one should bear in mind that a mixture of hydrogen and carbon monoxide has been being distributed in Europe and in the U.S.A without any major incident since the 19th century. In fact, it is not H_2 but CO to be considered the most dangerous between the two due to the danger of carbon monoxide poisoning. In France a hydrogen network currently in operation is about 170 km long, while the overall installed pipelines are more than 1500 km in Europe and longer than 700 km in North America.

Transporting hydrogen gas is very different from transporting methane gas. Given the same mass, the energy contained in hydrogen is 2–3 times higher than what is stored in methane. Since the pipeline transport speed is directly proportional to $\frac{1}{\sqrt{m}}$ (m being the molecular mass), it follows that hydrogen moves around 3 times faster than methane.

All in all, considering the numerous challenges in constructing, maintaining and managing hydrogen transport networks, a stand-alone hydrogen system maybe a preferable option for many applications. This can even become a realistic start-up development model during the initial period of transition from the current fossil fuel economy to a new one based on hydrogen. For many stationary and non-stationary uses, it can be more economically viable to use self-generated hydrogen than to purchase it from other industrial suppliers.

References

1. Agbossou K, Chahine R, Hamelin J et al (2001) Renewable energy systems based on hydrogen for remote applications. Journal of Power Sources 96:168–172
2. Blarke M B, Lund H (2008) The effectiveness of storage and relocation options in renewable energy systems. Renewable Energy 7 (33):1499–1507
3. Cox K E, Williamson K D (1977) Hydrogen: its technology and implications. (1) CRC Press, Cleveland
4. Keith G, Leighty W (2009) Transmitting 4000 MW of new windpower from N Dakota to Chicago: new HVDC electric lines or hydrogen pipeline. Proc. 14th World Hydrogen Energy Conference, Montreal, Canada

5. Ledjeff K (1990) New hydrogen appliances in Veziroğlu T N and Takahashi P K (Eds.) Hydrogen Energy Progress, VIII, vol. 3. Pergamon Press, New York, pp. 429–444
6. Ross D K (2006) Hydrogen storage: The major technological barrier to the development of hydrogen fuel cell cars. Vacuum 10 (80):1084–1089
7. Züttel A (2003) Materials for hydrogen storage. Materials today, September: 24–33

3

Electrolysis and Fuel Cells

Although hydrogen is one of the most commonly-found elements in the universe, it rarely exists as an independent molecule on our planet. Most of the time, it is bound to other elements or molecules to form compounds like water, carbohydrates, hydrocarbons and DNA acids. Obtaining hydrogen is not easy and usually requires a certain amount of energy to break the bonds connecting hydrogen to other elements. One process is water electrolysis in which electric energy is used to split water into hydrogen and oxygen. To regain the potential chemical energy stored in the hydrogen molecule, hydrogen and oxygen are combined to yield energy and water in the fuel cell which works in the reaction opposite to the electrolysis. This chapter discusses these two processes occurring in the electrolyser and in the fuel cell, two fundamental components of the solar hydrogen energy system.

3.1 Introduction

In order to be used for energy purposes, hydrogen must first be obtained then stored. To achieve this goal, a number of systems with different functioning technologies must be efficiently integrated together.

Hydrogen production[1] can be performed with water *electrolysis*, where electricity separate water molecules into hydrogen and oxygen in a device called *electrolyser*. The device which re combines hydrogen and oxygen in order to convert the chemical energy stored in them into electricity is called *fuel cell*. The reaction in the fuel cell is the same reaction occurring in the electrolyser but in the opposite direction.

Michael Faraday was one of the forerunners to start conducting a systematic study on electrolysis.

---

[1] The term *production* is not intended in the sense that hydrogen is *created*, but only that hydrogen is obtained in its simplest form (atomic or molecular) by processes that separate hydrogen from other compounds in which hydrogen is one of the constituents. The term will be anyhow used during the course of the book just as in common industrial practices.

Zini G., Tartarini P.: Solar Hydrogen Energy Systems. Science and Technology for the Hydrogen Economy. DOI 10.1007/978-88-470-1998-0_3, © Springer-Verlag Italia 2012

In 1832 he proposed the *two laws of electrolysis*:

- the quantity of the elements produced during electrolysis is directly proportional to the amount of the electricity passing through the electrolytic cell;
- with a given quantity of electricity, the amount of the elements produced is proportional to the equivalent weight[2] of the element.

The electrolyser and the fuel cell base their functions on these two laws.

3.2 Chemical Kinetics

A chemical reaction can be generally described with the following:

$$jA + kB \rightarrow lC + mD \tag{3.1}$$

where A, B, C and D are the reacting chemical species and j, k, l, m are the respective stoichiometric coefficients.

From the *law of mass action* (by Guldberg and Waage) at equilibrium and fixed temperature, the constant of reaction K is given by:

$$K = \frac{a_C^l \, a_D^m}{a_A^j \, a_B^k} \tag{3.2}$$

where a_i is the activity of the reacting substances, calculated as the concentration in the solution or the partial pressure at equilibrium. The activity can be given, for instance, by:

$$a_i = \frac{p_{i,eq}}{p_{stc}} \tag{3.3}$$

where $p_{i,eq}$ is the partial pressure of reactants i at equilibrium and p_{stc} is the pressure in standard conditions[3].

In case of an electrochemical reaction, the potential E of the cell in which the reaction takes place is given by the *Nernst's equation*:

$$E = E^0 - \frac{RT}{zF} \log Q \tag{3.4}$$

where Q is the reaction quotient, E^0 is the potential in standard conditions, R is the universal constant of gases, z is the number of electrons involved in the electrochemical reaction and F is the Faraday's constant.

---

[2] In chemistry, the *equivalent weight* (or *equivalent mass*) is defined as the quantity of mass of a substance able to supply or consume one mole of electrons in a redox reaction, or to generate one mole of H^+ ions by dissociation or one mole of OH^- ions in an acid-base reaction. It is calculated as the ratio between the molecular weight of the substance (expressed as g/mol) and its number of moles participating in the reaction. A *mole* is the quantity of a substance that contains the same amount of elementary entities as the number of atoms present in 12 g of C^{12}. Such number is known as the *Avogadro's number*, equal to 6.022×10^{23}.

[3] Standard conditions refer to a pressure of 0.1 MPa and a temperature at 25 °C. For a chemical element, the standard state is the condition it assumes at standard pressure and temperature.

According to *Le Chatelier's principle*, every chemical system reacts to an externally imposed modification to minimize the effect of the change. Therefore if there is a perturbation, the system will shift either towards the products or to the reactants to counteract the change. At equilibrium, Q equals K.

3.3 Thermodynamics

The enthalpy variation ΔH in a chemical reaction is defined as the difference between the sum of the enthalpy of formation of the products and that of the enthalpy of formation of the reactants:

$$\Delta H = \sum H_{f,products} - \sum H_{f,reactants}. \tag{3.5}$$

A chemical reaction is *exothermic* when thermal energy is released to the environment; in this case the enthalpy difference is negative. When the difference is positive, the reaction is *endothermic* and occurs only when it absorbs energy from the external environment.

The energy effectively available to generate work is what remains from ΔH after removing the product of temperature T and of entropy S. The resulting state function is the *Gibbs' free energy*. This function is important because in nature these transformations usually occur at a constant pressure and temperature rather than with a fixed volume. It is defined as:

$$G = H - TS. \tag{3.6}$$

The Gibbs' free energy determines if a chemical reaction will happen spontaneously at a given temperature, since a spontaneous reaction occurs only when the variation of free energy ΔG is negative. The infinitesimal difference of Gibbs' free energy can be expressed as:

$$dG = dH - T\,dS - S\,dT. \tag{3.7}$$

When T is constant:

$$dG = dH - T\,dS \tag{3.8}$$

with:

$$dH = dU + p\,dV + V\,dp \tag{3.9}$$

where U is the internal energy, p the pressure and V the volume.

If p is also constant:

$$dG = dU + p\,dV - T\,dS. \tag{3.10}$$

If a thermodynamic transformation happens between two infinitely close equilibrium states, the first principle of Thermodynamics can be described as[4]:

$$dU = \delta Q - \delta L. \tag{3.11}$$

---

[4] While the internal energy is an exact differential because it depends only on the initial and final states, Q and W are not state functions, therefore their integral depends on the cycle. For this reason the symbol δ is used instead of d to indicate that it is not about the exact differentials but rather the infinitesimal quantities of heat and work.

From which it is derived that:

$$dG = \delta Q - \delta L + p\,dV - T\,dS. \tag{3.12}$$

In an reversible transformation, δQ equals $T\,dS$, therefore the previous formula can be reduced to:

$$dG = -(\delta L - p\,dV). \tag{3.13}$$

For this reason, in electrochemical cells all the work that is not lost as volume change is available as electric work. In a reversible transformation, the work calculated as variation of Gibbs' free energy is ideal work.

The chemical reaction of water being split into hydrogen and oxygen contains the same element as water formation, except for the obvious fact that they occur in the opposite directions. Apart from the signs, the thermodynamics of both reactions are also the same.

3.4 Electrode Kinetics

The kinetics occurring at the electrodes of an electrolytic cell depend on the technology, the structure, the geometric layout of the cell, the type of the electrolyte used and other factors that can reduce the conversion efficiency. The efficiency reduction is called *polarisation* (or *over-potential, over-voltage*) and is attributed to the electromotive forces manifested during the cell function.

The phenomena of polarisation involve both the anode and the cathode. Depending on the direction of the reaction, polarisations tend to increase or lower the anode voltage where the oxidation reaction takes place and lower or increase the cathode voltage where the reduction reaction occurs. For this reason they tend to increase the electromotive force needed for electrolysis or lower the output voltage of the fuel cell.

3.4.1 Activation Polarisation

For a chemical reaction to occur, the reaction must be capable of overcoming the activation energy, namely the minimum energy required for the reaction to start. *Activation polarisation* η_{act} is the minimum voltage required between the electrodes of the cell to initiate the reactions. In an electrochemical reaction this voltage is in the range of 50–100 mV.

η_{act} is expressed by the *Tafel's equation* as:

$$\eta_{act} = \frac{R\,T}{\alpha z F} \log \frac{i_o}{i} \tag{3.14}$$

where α is the coefficient of the *charge transfer coefficient*, i_o is the density of the exchange current and i is the density of the current passing through the electrode surface. The charge transfer coefficient depends on the reaction mechanisms between the electrons and the catalysts and usually acquires a value between one and zero.

Some semi-empirical formulas are available to calculate the exchange current. The following equations have been proposed in related literature [2]:

$$i_{0,anode} = 5.5 \times 10^8 \left(\frac{p_{H_2}}{p_0}\right)\left(\frac{p_{H_2O}}{p_0}\right)\exp\left(\frac{-100 \times 10^5}{RT}\right) \qquad (3.15)$$

$$i_{0,cathode} = 7 \times 10^8 \left(\frac{p_{O_2}}{p_0}\right)^{0.25}\exp\left(\frac{-120 \times 10^3}{RT}\right) \qquad (3.16)$$

where p_0 is the value of the pressure in standard conditions and p_{H_2} and p_{O_2} are the partial pressures of hydrogen and oxygen.

3.4.2 Ohmic Polarisation

Ohmic losses are attributed to the electrode material resistance to the electron flow and the electrolyte resistance to the ion flow. Since most ohmic losses are caused by the resistance of the electrolyte, they can be reduced by drawing closer the two electrodes and by diminishing the thickness of the electrolyte. The losses caused by ohmic polarisation can be expressed by the equation:

$$\eta_{ohm} - IR \qquad (3.17)$$

where I is the current in the cell and R is the total resistance of the cell.

3.4.3 Concentration Polarisation

The transport phenomena of the mass of the reactants and the products at the entry and exit points of the cell can hinder the operation of the device if the flow rate is not sufficiently fast to maintain the current density in operation.

This *concentration polarisation* occurs when the current density is high. It is attributed to the slow diffusion of the reactants in the electrolyte with the creation of strong concentration gradient and subsequent voltage changes with respect to the ideal case with no polarisations.

From *Fick's first law*[5], the diffusive transport can be described as:

$$i - \frac{nFD(C_R - C_S)}{\delta} \qquad (3.18)$$

in which D is the reactants diffusion coefficient, C_B is their concentration in the electrolyte, C_S is the concentration on the electrode surface and δ is the thickness of the

[5] The *Fick's first law* states that the flux of molecules in a fluid occurs from high concentration areas to low-concentration regions. The diffusive flux J is given by: $J = -D\nabla\phi$, where D is the diffusion coefficient that depends on the size of the diffusing molecules, the temperature and the fluid viscosity, while ϕ is the spatial concentration of the molecules. The *Fick's second law* gives the change in time of the concentration when the molecules diffuse in a fluid as: $\frac{\partial\phi}{\partial t} = D\nabla^2\phi$.

diffusive layer. When C_S is close to zero, the maximum limit value of i_L can be calculated and it can be reached when the concentration of the reactants at the entry point is too low.

With a few passages from Equation (3.18):

$$\frac{C_S}{C_B} = 1 - \frac{i}{i_L}. \tag{3.19}$$

From the Nernst's equation at equilibrium (hence with no currents inside the cell):

$$E_{i=0} = E_0 + \frac{RT}{nF} \log C_B \tag{3.20}$$

that becomes, for values outside of equilibrium (non null currents):

$$E = E_0 + \frac{RT}{nF} \log C_S. \tag{3.21}$$

From the previous equations, the electrode voltage variation caused by the change of the concentration is expressed by:

$$\eta_{conc} = \Delta E = \frac{RT}{nF} \log\left(\frac{C_S}{C_B}\right) = \frac{RT}{nF} \log\left(1 - \frac{i}{i_L}\right). \tag{3.22}$$

3.4.4 Reaction Polarisation

The *reaction polarisation* occurs when the chemical reaction in the cell yields new chemical species or changes the equilibrium of the reaction. The variations in the concentrations of the reactants and the products during the cell operation therefore can reduce the conversion efficiency. The synthesis of water, for example, dilutes the solution itself and changes the electrolytes concentration on the electrodes surface.

3.4.5 Transfer Polarisation

The voltage variation occurring at the electrodes depends on the behaviour of the same electrodes. The over-potential required to drive the current flow between the electrodes is called *transfer polarisation* and can be expressed with an empirical, non-linear equation:

$$U = a + b \log i \tag{3.23}$$

where a e b are the coefficients determined by experiments.

3.4.6 Transport Phenomena

The irreversible energy losses occurring in the electrolytic solution are caused by the transport phenomena related to the exchange of heat, mass and electric charges. Other types of losses occur on the electrode surface when the speed of the reactions is not sufficiently fast.

Since the transport phenomena render the behaviour of an electrochemical reactor highly anisotropic, it is more feasible to adopt mathematical models considering that the laws of conservation of heat, mass and electric charges are being applied to the electrochemical reactor as an entire unity (*lump model*). This approximation can be legitimate and does not produce significant errors, since the behaviour is studied at a higher overall system level.

3.4.7 Influence of Temperature and Pressure on Polarisation Losses

A temperature increase can improve cell conductivity and reduce the losses related to ohmic polarisation. A higher temperature also improves its chemical kinetics and lowers the losses of activation polarisation. The undesirable collateral effect, however, is that high temperature not only causes the deterioration and the corrosion of the electrolyte but also creates problems of sintering and crystallisation of the catalysts.

The pressure increase at the cell entry point also boosts the partial pressure of the reactants and improves the solubility of the gas in the electrolyte, facilitating therefore the transport phenomena. Also in this case, against the positive effects, the pressure increase will significantly stress device materials.

If heating and pressurisation systems are required to improve the operation of the cell, the additional energy costs will tend to reduce the overall system efficiency. A careful cost/benefit analysis will be helpful in optimising the system performance.

3.5 Energy and Exergy of the Cell

If the pressure and the temperature in an electrochemical reaction are constant, the ideal maximum obtainable work (in the fuel cell) or the minimum required work (in the electrolyser) equals the variation of the Gibbs' free energy ΔG. In an electrochemical cell the ideal reversible work is an electric work, hence the equation:

$$W_{el} = \Delta G. \tag{3.24}$$

By convention this work is set positive for the electrolyser and negative for the fuel cell.

All energy sources are equivalent if viewed from the first principle of Thermodynamics. The *first principle efficiency* is defined as the ratio between the energy obtained from the system and the energy supplied to the system. Normally, this is the way efficiency coefficients for energy systems are computed. The output energy from the system and the corresponding input energy required are considered equivalent. But not all forms of energy are equivalent. For example, kinetic energy can be transformed into thermal energy that remains available only at the temperature of the environment. The two forms of energy therefore are not equivalent in terms of their quality, since kinetic energy is more usable than low temperature thermal energy.

For this reason, *Exergy* is defined as the maximum work obtainable from a heat source when the temperature $T > T_a$, where T_a is the environmental temperature. It can also be the minimum quantity of work needed to make an amount of heat available when $T < T_a$. To correctly take into consideration the difference of the qualities of the energy, the *second principle efficiency* (or *exergy efficiency*) is applied, which is defined as the ratio between the exergy provided by the system and the exergy supplied to the system.

In terms of the efficiency of the first principle of Thermodynamics, since the fuel exergetic power coincides with its lower heating value, the efficiency η_I of the first principle of Thermodynamics can get close to unity. If the cell works as an electrolyser:

$$\eta_I = \frac{\dot{m}_c H_i}{P} \tag{3.25}$$

where H_i is the lower heating value of the fuel. In case the cell works as a fuel cell:

$$\eta_I = \frac{P}{\dot{m}_c H_i}. \tag{3.26}$$

In terms of the efficiency of the second principle of Thermodynamics, since there is no intermediate conversion of the thermal energy, the electrochemical reactions can be reversible with exergetic efficiency η_{II} equal to one:

$$\eta_{II} = \frac{\dot{m}_c e_c}{P} \tag{3.27}$$

for a cell working as electrolyser, with P the power obtained from the electrochemical reaction, $\dot{m}_c$ the fuel mass capacity and e_c the fuel exergetic power. In case of the fuel cell:

$$\eta_{II} = \frac{P}{\dot{m}_c e_c}. \tag{3.28}$$

3.6 Electrolyser

3.6.1 Functioning

An electrolytic cell is a device where oxidation-reduction (redox) reactions occur to decompose chemical compounds by electric energy. In the cell, an electrolyte, present in the form of an acid (like HCl), a base (a hydroxide like NaOH) or a salt (like NaCl), reacts with the solvent (usually water) and splits into positive and negative ions (i.e. H^+, Na^+, OH^- or Cl^-). By connecting two electrodes immersed in the electrolytic cell to an outer electric circuit in which an electromotive force is applied, an electron flow is produced through the external circuit and corresponded by an ion flow in the cell internal electrolyte solution. The cathode is the electrode through which the electrons enter the cell and the reduction half-reaction occurs, while the anode is the electrode from which the electrons exit the cell and where the oxidation half-reaction takes place.

All chemical species are categorised with respect to their *reduction potential* (measured in V) that represents its propensity to be reduced or, equivalently, to gain electrons from another chemical species. The higher the reduction potential of a chemical species, the higher its tendency to gain electrons with respect to other species. A chemical species with a lower reduction potential will tend to lose electrons to species with higher reduction potential. The reduction potential is measured in standard conditions with respect to a reference *standard hydrogen electrode*[6] (SHE) with a potential conventionally set at 0 V.

A thermodynamic scale of reduction potentials records the values ΔV_{rid} of all the elements measured against the SHE, whose ΔV_{rid} is set at zero. As a reversible electrochemical reaction, the oxidation potential ΔV_{oss} compared to the corresponding reduction potential has the same absolute value but the opposite sign ($\Delta V_{oss} = -\Delta V_{rid}$). For example, potassium has a very low standard reduction potential ($\Delta V_{rid} = -2.92\,V$), therefore it has a very low tendency to gain electrons, but it has a very high standard oxidation potential ($\Delta V_{oss} = 2.92\,V$), meaning a high tendency to lose electrons.

Industrial electrolysers usually consist of more than one electrolytic cell. In order to increase productivity and to evenly distribute the voltage drop, these electrolysers are constructed by several cells connected in series, while the metallic separator between two cells can work as a bipolar plate and function as the anode on one side and the cathode on the other.

The efficiency of electrolysis is calculated as the ratio between the chemical energy contained in the yielded hydrogen and the electric power employed to the process.

3.6.2 Technology

3.6.2.1 Alkaline Electrolysers

Alkaline electrolysers (AE) take up a large portion of the commercial electrolyser market. They are constructed with materials resistant to the attack of potassium hydroxide (KOH) and are designed in a way to prevent electrolyte leakage. The anode is made of nickel while the cathode consists of nickel coated with platinum. The operating temperature is between 70–85 °C and the electric current density on the electrodes is around 6–10 kA/m^2 with an efficiency ratio between 75%–85%.

An alkaline electrolyser is composed by an electrolytic cell in which two electrodes are immersed in a water solution of potassium hydroxide KOH. The OH$^-$ ions are drawn to the anode to be oxidized (i.e. to lose electrons), releasing oxygen molecules, water and the electrons which will enter the external electric circuit according to the following reaction:

$$2\,OH^- \rightarrow \frac{1}{2}\,O_2 + H_2O + 2\,e^- \quad (anode). \tag{3.29}$$

The oxidation potential ΔV_{rid} of this reaction is -0.40 V. At the cathode the released electrons are not absorbed by reduction to the metallic potassium of the K$^+$ ions contained in the liquid solution, since such reaction, as previously mentioned,

---

[6] The SHE absolute potential is estimated with a *Born-Haber cycle* to be 4.44 ± 0.02 V at 298.15 K.

has a very low reduction potential ($\Delta V_{rid} = -2.92\,\text{V}$). The reaction taking place at the cathode is the reduction of the water itself:

$$2\,H_2O + 2\,e^- \rightarrow H_2 + 2\,OH^- \quad (\text{cathode}) \tag{3.30}$$

with a reduction potential of -0.83 V, still negative but higher than that of the K^+ ions. Therefore at the cathode the water undergoes reduction (i.e. gains electrons) and releases hydrogen molecules and OH^- ions. The overall energy absorbed by the electrolysis is calculated by this sum:

$$\Delta V_{oss} + \Delta V_{rid} = -0.40\text{V} + (-0.83\text{V}) = -1.23\text{V} \tag{3.31}$$

which equals the electromotive force required to trigger the two non-spontaneous reactions.

A porous diaphragm, permeable to OH^- ions and water but not to H_2 and O_2 gas, allows water and the ionic electric current to pass through while keeping the two gases from mixing with each other, so that the system can store the two gases separately. The overall reaction is expressed as:

$$H_2O \rightarrow H_2 + \frac{1}{2}\,O_2 \tag{3.32}$$

which gives the inversion of the spontaneous and exothermic combustion reaction of hydrogen and oxygen.

3.6.2.2 Solid Polymer (Polymeric Membrane) Electrolysers

In *Solid polymer*, or *Polymer Membrane* electrolysers (SPE, PME), the electrolyte is made of a solid polymer which makes the construction and the maintenance processes simpler and more convenient. When the polymer is saturated with water, it becomes acid and permeable to ions. The cells can also be built with very thin electrolytes and the solidity of the membrane permits the cell function at a pressure up to 4 MPa and at temperatures between 80–150 °C.

Since SPE electrolysers are built with rare and expensive materials (the cathode is made with porous carbon and the anode with porous titanium), they are able to yield good efficiencies and high current densities.

3.6.2.3 High-Temperature Electrolysers

High-temperature electrolysis (HTE) requires heat at a very high temperature (around 1000 °C) and an expensive production process, but it guarantees high performance with a solid and non-corrosive electrolyte (based on zirconium and yttrium oxides) capable of conducting oxygen ions. The cathode is made of nickel while the anode can be made by nickel, nickel oxides and lanthanum. The current density at the electrodes is around 3-5 kA/m^2 with a cell voltage in the range of 1.0–1.6 V, with a hypothetically achievable efficiency close to 95%.

3.6.3 Thermodynamics

Assuming that hydrogen and oxygen are ideal gases, that water is incompressible and that its liquid and gaseous phases are separated, then the variations of enthalpy, entropy and Gibbs' free energy of the electrolysis reaction can be calculated using pure hydrogen, oxygen and water in standard conditions.

As a consequence, the total enthalpy variation of the system during the electrolysis of water is calculated as the enthalpy difference between the products (hydrogen and oxygen) and the reactants (water):

$$\Delta H = \Delta H_{H_2} + \frac{1}{2}\Delta H_{O_2} - \Delta H_{H_2O} \tag{3.33}$$

from which the following can be derived:

$$\Delta H_x = c_{p,x}\left(T - T_{ref}\right) + \Delta H^0_{f,x}. \tag{3.34}$$

Similarly, the entropy variation ΔS is:

$$\Delta S = \Delta S_{H_2} + \frac{1}{2}\Delta S_{O_2} - \Delta S_{H_2O} \tag{3.35}$$

where:

$$\Delta S_x = c_{p,x}\ln\left(\frac{T}{T_{ref}}\right) - R\ln\left(\frac{p}{p_{ref}}\right) + \Delta S^0_{f,x} \quad \text{for } x = H_2, O_2 \tag{3.36}$$

$$\Delta S_{H_2O} = c_{p,H_2O}\ln\left(\frac{T}{T_{ref}}\right) + \Delta S^0_{f,H_2O}. \tag{3.37}$$

In all these equations:

- $c_{p,x}$ is the specific heat of the species x at a constant pressure (for hydrogen molecules it is 28.84 J/mol K, for oxygen molecules 29.37 J/mol K and for liquid water 75.39 J/mol K);
- ΔH_x is the enthalpy variation of the species x in J/mol;
- $\Delta H^0_{f,x}$ is the enthalpy of the formation of the species x in standard conditions (the formation enthalpies of hydrogen and oxygen molecules are both zero by definition);
- p is the pressure measured in Pa (1 atm = 101.325 Pa);
- R is the universal constant of the gas, which equals 8.314 J/mol K;
- ΔS_x is the entropy variation of the species x in J/mol K;
- $\Delta S^0_{f,z}$ is the entropy of the formation of the species x in standard conditions in J/mol K;
- T is the temperature in K.

In standard conditions, water-splitting is not a spontaneous reaction and therefore requires a positive Gibbs' free energy variation, which in standard conditions is expressed as $\Delta G^0_{s,H_2O} = 237$ kJ/mol. The opposite reaction instead is spontaneous with a negative free energy measured as $\Delta G^0_{f,H_2O} = -237$ kJ/mol. In standard conditions, the enthalpy generated by water-splitting is 286 kJ/mol and therefore the enthalpy of water formation equals -286 kJ/mol, with $\Delta S^0_{f,H_2O} = 0.16433$ kJ/mol K.

As previously mentioned, the electrolysis reaction is not spontaneous ($\Delta G>0$) and requires an external electric work given by:

$$L_{el} = \Delta G = qE \tag{3.38}$$

where q is the electron charge transferred in the external circuit of the cell (in C/mol). One mole of water produces a charge of q composed by z moles of electrons as per Faraday's laws:

$$L_{el} = \Delta G = qE = zFE \tag{3.39}$$

where $z = 2$ and F is the Faraday's constant.

In an electrochemical reaction, E is called *reversible cell voltage U_{rev}*, therefore:

$$U_{rev} = \frac{\Delta G}{zF}. \tag{3.40}$$

In addition, the *thermo-neutral voltage U_{th}* is defined as the electromotive force required for the reaction to progress at a constant temperature. The thermo-neutral voltage is linked to the enthalpy variation by:

$$U_{th} = \frac{\Delta H}{zF} \tag{3.41}$$

in which $U_{rev} = 1.229$ V and $U_{th} = 1.482$ V in standard conditions.

An increase in temperature would decrease the reversible voltage (at 80 °C and 1 bar: $U_{rev} = 1.184$ V) while the thermo-neutral voltage would remain almost unchanged (at 80 °C and 1 bar: $U_{th} = 1.473$ V). A rise in the system pressure would cause a slight increase in the reversible voltage (at 25 °C and 30 bar: $U_{rev} = 1.295$ V) while the thermo-neutral voltage would stay the same.

3.6.4 Mathematical Model

Since it is considerably difficult to describe the relationship between the current and the voltage in an electrochemical reactor, the mathematical models that have been developed are empirical in nature. In this context, the *I-U* characteristic curve of a PEM electrochemical cell can be described using the following model [8]:

$$U_{el} = U_{el,0} + C_{1,el}\, T_{el} + C_{2,el} \log\left(\frac{I_{el}}{I_{el,0}}\right) + \frac{R_{el} I_{el}}{T_{el}} \tag{3.42}$$

where $U_{el,0}$, $C_{1,el}$, $C_{2,el}$, $I_{el,0}$ and R_{el} are the parameters obtained from experiments, T_{el} is the cell temperature, U_{el} is the voltage measured at the electrodes and I_{el} is the current flowing in the electrolyser (see Figure 3.1). The first two terms represent the thermo-neutral voltage of all the cells connected in series in ideal condition, calculated as the sum of the voltage from each individual cell. The third term includes the

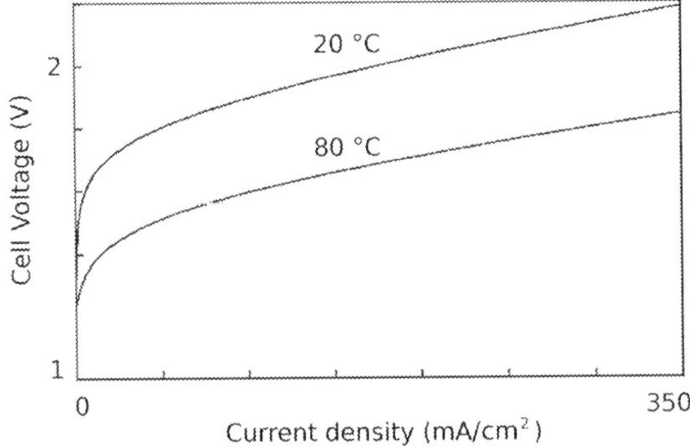

Fig. 3.1. Current-voltage curves in an electrolytic cell at different temperatures

losses from the activation polarisation while the forth term stands for the ohmic polarisation losses, with the losses from the concentration polarisation implicitly included in the parameter that have been determined empirically.

The operating point of the electrolyser is given by the following non-linear system, with a solution that can be obtained through numerical calculation:

$$\begin{cases} P_{el} = U_{el} I_{el} \\ U_{el} = U_{el,o} + C_{1,el} T_{el} + C_{2,el} \ln\left(\frac{I_{el}}{I_{el,0}}\right) + \frac{R_{el} I_{el}}{T_{el}} \end{cases} \tag{3.43}$$

where P_{el} is the electric power supplied to the electrolyser.

In accordance with the Faraday's laws, the flow rate of the hydrogen produced is proportional to the electric current passing through the external circuit:

$$\dot{n}_{H_2} = \eta_F \frac{N_c I_{el}}{z F} \tag{3.44}$$

where $\dot{n}_{H_2}$ is the molar flow rate of the hydrogen produced (in mol/s) and N_c is the number of the electrolytic cells connected in series in the electrolyser. The molar flow rate of the water supplied and that of the oxygen produced can be described by the following:

$$\dot{n}_{H_2O} = \frac{1}{2}\dot{n}_{O_2} = \dot{n}_{H_2} \tag{3.45}$$

where η_F is the Faraday performance of the electrolyser, defined as the ratio between the number of moles effectively produced by the electrolyser and the maximum number that could be theoretically obtained.

The *electric efficiency* η_e is defined as the ratio between the thermo-neutral voltage sum of all the cells in the electrolyser and the voltage of the electrodes U_{el}:

$$\eta_e = \frac{N_c U_{th}}{U_{el}}. \tag{3.46}$$

Normal electrolyser efficiencies are around 70%.

3.6.5 Thermal Model

The cell temperature, which is the same as that of the electrolyte, has a great influence on the *I-U* characteristics and on the device operation. The thermal balance of the electrolyser can be expressed by the following equation:

$$\dot{Q}_{gen} = \dot{Q}_{store} + \dot{Q}_{loss} + \dot{Q}_{cool} \tag{3.47}$$

in which $\dot{Q}_{gen}$ is the thermal power generated during operations, $\dot{Q}_{store}$ is the thermal power stored in the system mass, $\dot{Q}_{loss}$ is the thermal power released to the environment and $\dot{Q}_{cool}$ is the thermal power absorbed by the cooling system.

The cooling of the electrolyser is usually achieved by increasing the amount of water supply to as much as two times the stoichiometric quantity required for the electrolysis reaction. The excess water absorbs the heat in a way similar to how a heat exchanger functions, in which a flow of cooling water ($\dot{n}_{H_2O,cool} = \dot{n}_{H_2O}$) passes through the system and the released heat increases the temperature $T_{cw,in}$ of the water at the entry point up to the temperature $T_{cw,out}$ at the outlet.

The following relationships can be established:

$$\dot{Q}_{gen} = (U_{el} - N_c U_{th}) I_{el} = U_{el} I_{el} (1 - \eta_e) \tag{3.48}$$

$$\dot{Q}_{store} = C_t \frac{dT}{dt} \tag{3.49}$$

$$\dot{Q}_{loss} = \frac{(T - T_a)}{R_t} \tag{3.50}$$

$$\dot{Q}_{cool} = C_{cw} (T_{cw,in} - T_{cw,out}) = UA_{hx} \times LMTD \tag{3.51}$$

in which:

- C_t is the thermal capacity of the electrolyser measured in J/K;
- T is the electrolyte temperature;
- t is the time;
- T_a is the ambient temperature;
- R_t is the thermal resistance of the electrolyser in K/W;
- $T_{cw,in}$ and $T_{cw,out}$ are the temperatures of the cooling water at the entry and the exit points of the electrolyser respectively;
- UA_{hx} is the product (in W/K) of the thermal exchange coefficient and the surface of the equivalent heat exchanger;
- $LMTD$ is the log mean temperature difference of the heat exchanger;
- $C_{cw} = (\dot{n}_{H_2O,cool} \times c_{p,H_2O})$ is the thermal capacity of the flow of the cooling water, in which $\dot{n}_{H_2O,cool}$ is the molar flow rate of the cooling water (in mol/s) and c_{p,H_2O} is the molar specific heat of the water (in J/mol K) at a fixed pressure.

The thermal exchange coefficient can be written as:

$$UA_{hx} = a_{cond} + b_{conv} I_{el} \tag{3.52}$$

where a_{cond} (in W/K) is the reciprocal of the thermal resistance of the electrolyser R_t and b_{conv} (in W/K A) is a constant that depends on the convection coefficient between the electrodes and the electrolyte. UA_{hx} is directly proportional to the electric current density passing through the electrolyser I_{el}, because the density increase also augments the internal gas production which in turn speeds up the convective exchanges caused by the fluid remixing from the bubbles formed.

The log mean temperature difference $LMTD$ by definition equals:

$$LMTD = \frac{(T - T_{cw,i}) - (T - T_{cw,o})}{\ln\left[(T - T_{cw,i})/(T - T_{cw,o})\right]}. \tag{3.53}$$

Assuming the temperature of the electrolyte T is constant, the previous equation becomes:

$$T_{cw,o} = T_{cw,i} + (T - T_{cw,i})\left[1 - \exp\left(-\frac{UA_{hx}}{C_{cw}}\right)\right]. \tag{3.54}$$

Finally, combining the thermal equilibrium equations, the result is the differential, linear, non-homogeneous equation of the first order:

$$\frac{dT}{dt} + at - b = 0. \tag{3.55}$$

Its solution is:

$$T = \left(T_{in} - \frac{b}{a}\right)\exp(-at) + \frac{b}{a} \tag{3.56}$$

in which:

$$a = \frac{1}{\tau_t} + \frac{C_{cw}}{C_t}\left[1 - \exp\left(-\frac{UA_{hx}}{C_{cw}}\right)\right] \tag{3.57}$$

$$b = \frac{U_{el}I_{el}(1 - \eta_e)}{C_t} + \frac{T_a}{\tau_t} + \frac{C_{cw}T_{cw,i}}{C_t}\left[1 - \exp\left(-\frac{UA_{hx}}{C_{cw}}\right)\right]. \tag{3.58}$$

In the above equations, T_{in} is the initial temperature of the cell which equals the ambient temperature when the operation starts, whereas τ_t is the electrolyser constant of time in seconds, the same as the product R_tC_t. The temperature of the hydrogen and the oxygen at the outlet is considered the same as that of the electrolyte.

3.7 Fuel Cell

3.7.1 Functioning

Fuel cells (FC) are electrochemical devices where the spontaneous combustion of hydrogen and oxygen occurs:

$$H_2 + \frac{1}{2}O_2 \rightarrow H_2O + \text{electrons} + \text{heat} \tag{3.59}$$

in which hydrogen (as the fuel) and oxygen (as the oxidiser) both enter in the gaseous state. The reaction yields water, electricity and heat without generating pollutants.

Most types of fuel cells do not generate greenhouse gases either. A fuel cell is therefore a combined heat and power system which yields both electricity and thermal energy.

In the fuel cell, hydrogen diffuses at the anode by penetrating through the electrode pores. By the reaction with the catalyst, the adsorbed hydrogen is ionized and then passes through the solution where it releases an electron at the electrode. The reaction is expressed as:

$$\frac{1}{2}\,H_2 \to [H] \to H^+ + e^- \tag{3.60}$$

in which the square parenthesis indicate the midway process of hydrogen adsorption inside the electrode porous structure. The hydrogen ions pass through the electrolyte and reach the cathode, where they recombine with oxygen molecules and with the electrons entering the cathode (which come from the anode after passing through the external circuit) to form water molecules. The heat generated by the cell internal reaction can be retrieved by adopting suitable water or air cooling systems.

A fuel cell is assembled by several sub-systems responsible for managing the whole process of the hydrogen energy conversion. For instance:

- a control system which regulates the input and output flows with pumping devices. Such system also needs to ensure that the flows reach the correct temperature, pressure and humidity level as requested by the technical specifications of the cell;
- a thermal control system which optimizes the operation temperature, dissipates and retrieves the excess heat with the possibility of CHP applications;
- a system which conditions the electric output.

Apart from being extremely useful and adaptable, fuel cells also own many other advantages, such as:

- high conversion rates;
- efficient CHP uses;
- low operating pressures and temperatures that range from 80 °C to 1000 °C, lower than the 2300 °C of internal combustion engines;
- scalability of the system, so that the performance efficiency is independent from the size of the plant (since the number of the cells can be increased to enlarge capacity without compromising the performance);
- reduced environmental impact together with the possibility to be used along with different fuels;
- performance efficiency independent from load variations;
- high reaction speeds in response to changing loads;
- ease of construction with no moving parts.

Against these benefits, however, there remains the disadvantage of using rare materials, such as platinum as the catalyst. In addition, the service life of the cells is relatively short, which is estimated around 10000–20000 h with current technologies. Hopefully the technological advances in the near future will reduce the impacts of, or eliminate altogether, these shortcomings.

3.7.2 Technology

Fuel cells are categorized according to the types of electrolyte they use and their ranges of operation temperature. The main categories are:

- Alkaline Fuel Cell (AFC);
- Phosphoric Acid Fuel Cell (PAFC);
- Proton Exchange Membrane Fuel Cell or Polymeric Electrolyte Membrane Fuel Cell (PEMFC, PEM);
- Molten Carbonate Fuel Cell (MCFC);
- Solid Oxide Fuel Cell (SOFC).

3.7.2.1 Alkaline Fuel Cell

In an alkaline fuel cell the electrolyte is generally potassium hydroxide in a water solution with the operating temperature between 60–100 °C. The electrodes are made of porous carbon and the catalysts are made of nickel and platinum.

Just as in an electrolytic cell, in a fuel cell the redox reactions also take place at the electrodes: the electrons exit from the cell from the anode through the external load and re-enter in the cell at the cathode.

At the porous anode, hydrogen reacts with OH^- ions in the solution to form water and release electrons as per the following reaction:

$$H_2 + 2\,OH^- \rightarrow 2\,H_2O + 2\,e^- \tag{3.61}$$

with an oxidation potential of: $\Delta U_{oss} = -\Delta U_{rid} = 0.83$ V.

At the cathode, the electrons re-enter the external load and react with oxygen and water to produce OH^- ions:

$$\frac{1}{2}\,O_2 + H_2O + 2\,e^- \rightarrow 2\,OH^- \tag{3.62}$$

with a reduction voltage of $\Delta U_{rid} = -\Delta U_{oss} = 0.40$ V. The OH^- ions inside the electrolytic solution move towards the anode to close the internal circuit. The voltage therefore can be calculated by the sum of $\Delta U_{oss} + \Delta U_{rid}$ of the two spontaneous reactions, which equals $0.83 + 0.40 = 1.23$ V. Depending on the cell operating temperature, water is released either in liquid or vapour form as the only emission from the cell.

The strengths of this type of cell are its fast start-up speed and high efficiency but the cell has a low tolerance for carbon oxides, therefore AFC needs to be powered only by pure oxygen, which in turn restricts its wide-spread adoption.

3.7.2.2 Phosphoric Acid Fuel Cell

Phosphoric acid fuel cells use phosphoric acid as the electrolyte with an operation temperature falling in the range of 160 to 220 °C. The electrodes are made of gold, titanium and carbon with platinum as a catalyst to avoid corrosion. These cells have a good tolerance for CO_2 and therefore ambient air can be used to react with hydrogen.

It is impossible however to use traditional fossil fuels as the platinum electrodes can suffer from deterioration when it comes in contact with CO.

The reaction at the anode can be expressed as:

$$2\,H_2 \rightarrow 4\,H^+ + 4\,e^- \tag{3.63}$$

while the reaction at the cathode is:

$$O_2 + 4\,H^+ + 4\,e^- \rightarrow 2\,H_2O. \tag{3.64}$$

Working at a temperature between 160–220 °C, these cells can be used in small and medium-sized plants as well as in residential heating and hot water production, making them therefore suitable for CHP applications.

As PAFCs are CO_2 tolerant, they can be powered not only by pure or mixed hydrogen but also by hydrocarbons and alcohols. The downside however lies in its need to use an external reformer, which further complicates the cell operation and also increases the already elevated construction costs due to the necessity to use platinum as the catalyst.

Usually the dimension of a commercial plant of this type of cell is between 100 to 200 kW, but plants with a scale of several MW have been built in Japan and in the U.S.A to supply power to facilities like hospitals, offices, schools and airport terminals.

3.7.2.3 Polymeric Electrolyte Membrane Fuel Cell

Polymeric electrolyte membrane fuel cells have two porous carbon electrodes separated by a polymeric membrane of chlorinated sulphuric acid and operate at temperatures between 80–120 °C. The conduction ions are H^+ and the cell has the same electrochemical reactions as in PAFC. Ambient air can replace pure hydrogen as the fuel, provided that no CO residues are present to cause permanent damages to the catalyst.

PEM cells are highly reliable because of their non-corrosive polymeric electrolyte, the tolerance for carbon dioxide and the reduced thermal and mechanical solicitations thanks to the relatively low operating temperatures. The cells are also suitable for non-stationary applications as they are simple to install and easy to manage and maintain. Furthermore, the high current density (around 2 A/cm^2) allows the cell to become compact yet light, two very much appreciated qualities especially in automotive uses, not to mention its fast start-up speed and quick reaction to load variations. In stationary uses, PEMs can generate electricity and function as a power back-up device in systems with a capacity up to 200 kW. Another advantage is the relatively low construction costs of the cell, made possible thanks to the process of standardisation and the economies of scale accomplished from the numerous applications.

3.7.2.4 Molten Carbonate Fuel Cell

The electrolyte in *molten carbonate fuel cells* is a carbonated alkaline liquid solution (mainly KCO$_3$) and the cell operation temperature is between 600–850 °C.

The reactions at the anode are:

$$3\,H_2 + 3\,CO \rightarrow 3\,H_2O + 3\,CO_2 + 6\,e^- \tag{3.65}$$

$$CO + CO_2 \rightarrow 2\,CO_2 + 2\,e^-. \tag{3.66}$$

While at the cathode the reaction is:

$$2\,O_2 + 4\,CO_2 + 8\,e^- \rightarrow 4\,CO_3^- \tag{3.67}$$

in which the conducting ions are CO_3^-. The reactions at the electrodes correspond to the combustion of H_2 and CO with the production of CO_2 at the anode. Thanks to this feature, the cell can avoid having separate supplies of gases and the plant construction is further simplified.

MCFCs have very high conversion ratios and are suitable to be used in CHP applications combined with reformer systems. However, the battery has to sustain a considerable thermal stress that can reduces its reliability and service life. At the moment the cell is not suitable for non-stationary uses, but its ability to be powered by different types of fuels makes it extremely flexible and more commercially popular in stationary uses. Moreover, its high operating temperatures can reduce the use of catalysts and therefore cut down production costs.

3.7.2.5 Solid Oxide Fuel Cell

Solid oxide fuel cells use a type of solid and ceramic electrolyte composed by zirconium oxide (ZrO_2) stabilized with yttrium oxide (Y_2O_3), which allows the cell to operate at high temperatures (between 800 to 1000 °C). For this reason the cell no longer needs to use catalysts. At the same time, the cell can be powered by a mixture of hydrogen, carbon oxide and hydrocarbons.

The anodic reactions are:

$$H_2 + O^- \rightarrow H_2O + 2\,e^- \tag{3.68}$$

$$CO + O^- \rightarrow CO_2 + 2\,e^- \tag{3.69}$$

$$CH_4 + 4O^- \rightarrow 2\,H_2O + CO_2 + 8\,e^- \tag{3.70}$$

while the cathode reaction is:

$$3\,O_2 + 12\,e^- \rightarrow 6\,O_2^- \tag{3.71}$$

in which O^- is the conducting ion.

These reactions correspond to the combustion of hydrogen, oxygen, carbon oxide and methane. Unfortunately, the presence of carbon signifies the emission of the greenhouse gas, carbon dioxide.

Since it is difficult to produce large electrolyte foils that are thin and resistant at the same time, this type of cells are constructed with a particular geometry disposition, for example in a tubular shape. SOFC plants are usually large and complex and, at least for the time being, still not suitable for small facilities or non-stationary uses.

The cell start-up time to reach the operation temperature is longer compared to other types of fuel cells, but this doesn't prevent the device to be used efficiently in combined heat and power applications. The considerable thermal solicitations, however, can reduce the cell service life and reliability.

3.7.3 Thermodynamics

Compared to an electrolytic cell, a fuel cell has the exact opposite thermodynamics and a reverse chemical reaction direction. While the reaction in an electrolytic cell is not spontaneous, the energy conversion in a fuel cell is spontaneous with a negative Gibbs energy variation.

In standard conditions $\Delta G = -237$ kJ/mol, while in non-standard conditions it is calculated as:

$$\Delta G = \Delta H - T \Delta S \tag{3.72}$$

where the enthalpy change ΔH and the entropy change ΔS are obtained by:

$$\Delta H = \Delta H_{H_2O} - \frac{1}{2} \Delta H_{O_2} - \Delta H_{H_2} \tag{3.73}$$

and by:

$$\Delta S = \Delta S_{H_2O} - \frac{1}{2} \Delta S_{O_2} - \Delta S_{H_2}. \tag{3.74}$$

Every cell produces a maximum ideal electric work:

$$W = \Delta G = qE = z\text{F}E \tag{3.75}$$

where z is the number of the electrons involved in the reaction, F is the Faraday's constant and E is the electromotive force generated by the cell, also called the *reversible voltage* of the cell U_{rev} as the maximum theoretically obtainable voltage in an open circuit.

From the previous equation, the reversible voltage of the cell can be calculated by:

$$U_{rev} = \frac{\Delta G}{z\text{F}}. \tag{3.76}$$

Similarly, the total energy supplied by the cell ΔH is linked to the thermo-neutral voltage U_{th} through:

$$U_{th} = \frac{\Delta H}{z\text{F}}. \tag{3.77}$$

In standard conditions U_{rev} equals 1.229 V, while in non-standard conditions the reversible voltage is obtained from the value of ΔG calculated in operating conditions.

In any case, it is possible to analyse the effects that temperature and pressure have on the reversible voltage U_{rev} with the following derivatives.

$$\left(\frac{dU_{rev}}{dT} \right)_{p=cost} = \frac{\Delta S}{z\text{F}} \tag{3.78}$$

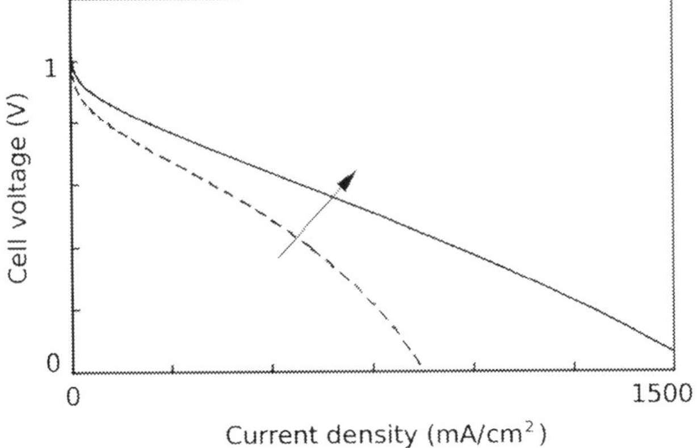

Fig. 3.2. Influence of temperature and pressure on the I-U characteristics of the fuel cell

$$\left(\frac{dU_{rev}}{dp}\right)_{T=cost} = -\frac{\Delta V}{z\mathrm{F}} \tag{3.79}$$

where ΔV represents the change in volume.

When the concentrations of the reactants are different from the stoichiometric value (i.e. excess of hydrogen or oxygen), U_{rev} changes according to the Nernst's equation:

$$U_{rev} = U_{rev}^0 + \frac{R\,T}{z\mathrm{F}} \log\left(\frac{1}{[\mathrm{H_2}]\sqrt{[\mathrm{O_2}]}}\right) \tag{3.80}$$

in which $[\mathrm{H_2}]$ and $[\mathrm{O_2}]$ are the concentrations of the hydrogen and oxygen molecules.

Since in the same reaction the volume change is negative (passing from gas to liquid phase), an increase in pressure can also produce some positive effects, such as the rise of reversible voltage which facilitates the transport phenomenon, the increase of gas solubility in the electrolyte and the reduction of material loss caused by evaporation. However, higher pressure also entails the collateral negative effect of subjecting the material to higher mechanical stresses. Temperature has also a beneficial effect on cell efficiency; the ohmic, activation and concentration polarisations are reduced by an increase in cell operating temperatures. The effect of pressure and temperature on cell performance is shown in Figure 3.2.

From the Faraday's laws it is possible to derive the relationship between the electric current and the molar flow rate of the hydrogen consumed by the cell in the ideal condition (i.e. in stoichiometric conditions), $\dot{n}_{\mathrm{H_2},id}$:

$$\dot{n}_{\mathrm{H_2},id} = \frac{I_{fc}}{z\mathrm{F}} \tag{3.81}$$

in which I_{fc} is the current measured at the electrodes assuming that the device only has a single cell.

Because of the losses caused by polarisations, the electrochemical conversion tends to slow down; not all the injected hydrogen reacts completely and the cell consumes a higher quantity of hydrogen than in ideal conditions. In order to quantify this consumption excess in real-life conditions, the Faraday's performance can be calculated as:

$$\eta_F = \frac{\dot{n}_{H_2,id}}{\dot{n}_{H_2}} \tag{3.82}$$

where $\dot{n}_{H_2}$ is the actual molar flow rate of the hydrogen required to generate the desired current.

As the result, the hydrogen consumed in a fuel cell can be calculated as:

$$\dot{n}_{H_2} = \frac{N_c I_{fc}}{\eta_F z F}. \tag{3.83}$$

In the same way, the molar flow rates of the hydrogen required and the water produced are:

$$\dot{n}_{H_2} = \dot{n}_{H_2O} = \frac{1}{2} \dot{n}_{O_2}. \tag{3.84}$$

For a hydrogen/air fuel cell, since oxygen constitutes 21% of the mass of the air, the amount of the air required for the cell to function is about 4.76 times the quantity of the pure oxygen.

The electric efficiency η_e of a fuel cell can be defined as:

$$\eta_e = \frac{U_{fc}}{N_c U_{th}} \tag{3.85}$$

where U_{fc} is the output voltage of the cell electrodes.

From Faraday's laws, the maximum current I_{max} yielded by the fuel cell is given by:

$$I_{max} = n F \frac{df}{dt} \tag{3.86}$$

in which $\frac{df}{dt}$, expressed in mol/s, is the theoretical maximum consumption speed of the reactants. The real actual current is given by a similar equation as Equation (3.86) with $\frac{df}{dt}$ computed for the actual velocities of the reacting species.

If the molar inflow of hydrogen is $\dot{n}_{H_2,in}$ and $\dot{n}_{H_2,used}$, the actual molar flow of hydrogen actually used (in mol/s), η_e becomes:

$$\eta_e = \frac{I_{real}}{I_{max}} = \frac{\dot{n}_{H_2,used}}{\dot{n}_{H_2,in}}. \tag{3.87}$$

For a cell working with excess hydrogen, the electric current performance is lower than unity. The fuel cells consuming all the hydrogen inflow (*dead-end cells*) (without any hydrogen leakage) can reach an efficiency of virtually 100%.

3.7.4 Mathematical Model

The polarisations of the fuel cell always increase the voltage at anode and lower that of the cathode, as expressed by the following:

$$U_{anode} = U_{rev,anode} + |\eta_{anode}| \tag{3.88}$$

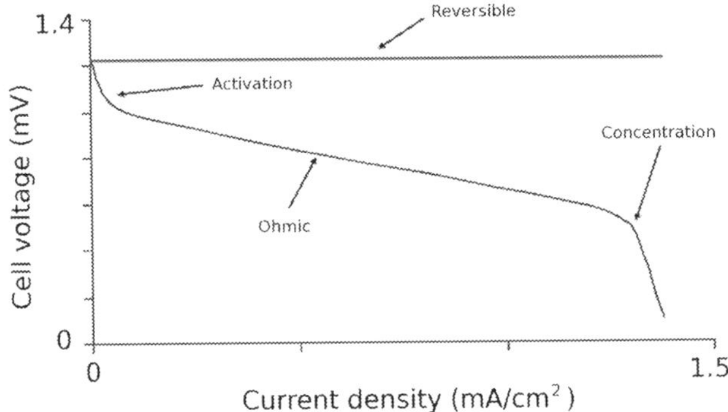

Fig. 3.3. *I-U* curve in a PEM fuel cell with the polarisation effects on the reversible voltage

$$U_{cathode} = U_{rev,cathode} - |\eta_{cathode}| \tag{3.89}$$

in which η_{anode} and $\eta_{cathode}$ represent the sum of the concentration and activation polarisations related to the anode and the cathode respectively. The cell voltage therefore can be expressed by:

$$U_{cell} = U_{cathode} - U_{anode} - RI_{fc} = U_{rev} - \eta_{conc} - \eta_{act} - \eta_{ohm}. \tag{3.90}$$

The resulted voltage-current density which has taken into account the effects of the polarisations is shown in Figure 3.3.

The non-linear behaviour of the fuel cell can be accurately modelled through experiments. The *I-U* curve of the cell can be described by the parameters that are derived by measuring the cell output in different operating conditions and, therefore, depend on the technology of the cell itself. The model for a PEM stack is expressed as, for example:

$$U_{fc} = U_{fc,0} + C_{1,fc}\, T_{fc} + C_{2,fc} \ln\left(\frac{I_{fc}}{I_{fc,0}}\right) + \frac{R_{fc} I_{fc}}{T_{fc}} \tag{3.91}$$

in which $U_{fc,0}, C_{1,fc}, C_{2,fc}, I_{fc,0}$ and R_{fc} are the parameters obtained experimentally, T_{fc} is the cell temperature ranging from 70–80 °C, U_{fc} is the stack voltage and I_{fc} is the current generated in the fuel cell.

The first two terms represent the thermo-neutral voltage of a series of cells in ideal conditions; the third term includes the losses caused by activation polarisation while the fourth contains the losses of ohmic polarisation. Same as in the model of the electrolyser, in a fuel cell the losses of concentration polarisation are determined empirically.

The operating point of the electrolyser is determined by solving the following non-linear system.

$$\begin{cases} P_{fc} - U_{fc} I_{fc} \\ U_{fc} = U_{fc,0} + C_{1,fc}\, T_{fc} + C_{2,fc} \ln\left(\frac{I_{fc}}{I_{fc,0}}\right) + \frac{R_{fc} I_{fc}}{T_{fc}} \end{cases} \tag{3.92}$$

where P_{fc} is the electric power provided by the fuel cell.

Library
University of Wisconsin

From the above system the value of I_{fc} can be obtained, which also permits the calculation of the consumed hydrogen flow rate $\dot{n}_{H_2}$ from the previous equations.

3.7.5 Thermal Model

The cell operating temperature needs to be controlled by an auxiliary cooling system, usually with the injection of water, so that the temperature of the polymeric electrolyte can remain constant in order to optimize the cell performance.

The cell stack can be heated up quite rapidly due its low thermal capacity. Without committing major approximation errors, the stack temperature can be considered as constant and equal to its nominal value during the whole operation, excluding the transients when the device is turned on and off.

References

1. Aylward G, Findlay T (1994) SI Chemical Data, 3rd ed. Wiley and Sons, New York
2. Costamagna P, Selimovic A, Del Borghi M, Agnew G (2004) Electrochemical model of the integrated planar solid oxide fuel cell (IP-SOFC). Chemical Engineering Journal 102:61–69
3. Incropera F P, DeWitt D P (1990) Fundamentals of Heat and Mass Transfer, 3rd ed. John Wiley & Sons, New York
4. Kordesch K, Simader G (1996) Fuel cells and their applications, 3rd ed. VCH Publisher Inc., Cambridge
5. Kothari R, Buddhi D, Sawhney R L (2005) Study of the effect of temperature of the electrolytes on the rate of production of hydrogen. Int. J. Hydrogen Energy 30:251–263
6. Leroy R L, Stuart A K (1978) Unipolar water electrolyzers. A competitive technology. Proc. 2nd WEHC, Zürich, Switzerland, pp. 359–375
7. Roušar I (1989) Fundamentals of electrochemical reactors. In: Ismail M I (ed.) Electrochemicals reactors: their science and technology, Part A. Elsevier Science, Amsterdam
8. Ulleberg Ø (2003) Modeling of advanced electrolyzers: a system simulation approach. Int. J. Hydrogen Energy 28:21–33
9. Williams M C, Strakey J P, Singhal S C (2004) U.S. distributed generation fuel cell program. J. Power Sources 131:79–85

4

Solar Radiation and Photovoltaic Conversion

Photovoltaic technology is able to collect and convert solar radiation photons to electric energy which then becomes readily available to supply a load or, when produced in surplus, to be stored for future uses. The solar radiation reaching the surface of the Earth varies due to meteorological conditions and the period of the year. Energy storage therefore becomes the key to render this type of renewable energy more reliable and more predictable.

4.1 Solar Radiation

Our closest star, the *Sun*, emits a continuous flow of electromagnetic energy generated by its internal nuclear fusion processes in which two hydrogen nuclei combine to form an atom of helium. Since the nuclear mass of the helium is smaller than the sum of the two diatomic hydrogen nuclei, the difference of the nuclear mass is converted into energy according to Einstein's law:

$$E = mc^2. \tag{4.1}$$

The Sun is a classified as a G2V yellow dwarf (according to the *Hertzsprung-Russell diagram*[1]), since most of the visible radiation is emitted in yellow-green spectrum. Its diameter is around 1.39×10^6 km. Nearly 75% of its mass consists of hydrogen while the rest is composed mostly by helium and less than 2% by other elements. Its surface temperature is around 5750 K.

Solar irradiance is the electromagnetic power the Earth receives from the Sun per unit surface (expressed in kW/m^2). From the observation of the NASA satellites, the total irradiance emitted by the Sun is measured at 3.8×10^{23} kW, out of which the Earth receives 173×10^{12} kW from an average distance of 6371 km. *Solar radiation* is expressed in kWh/m^2 and represents the electromagnetic energy per unit surface

---

[1] The *Hertzsprung-Russell diagram* is a graph showing the relationship between stars' absolute magnitudes, spectral classes, colours and temperatures.

Zini G., Tartarini P.: Solar Hydrogen Energy Systems. Science and Technology for the Hydrogen Economy. DOI 10.1007/978-88-470-1998-0_4, © Springer-Verlag Italia 2012

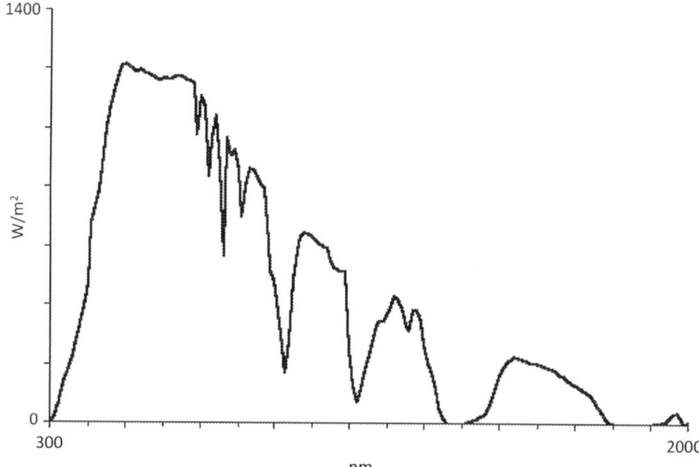

Fig. 4.1. Solar radiation spectrum at sea level with the bands of absorption caused by O_3, H_2O and CO_2

The amount of solar irradiance incident on a surface perpendicular to the Sun's rays and outside the Earth's atmosphere is called *solar constant*, which is in the range of 1367-1371 W/m^2 with a variation due to the elliptical orbit of the Earth and the intensity of the solar surface activity.

The intensity of the solar radiation absorbed by the terrestrial surface varies depending on the geographical coordinates, the day of the year, the different positions of the Sun across the sky during the course of the day, the mass of the air through which the radiation travels and the local meteorological conditions.

Solar radiation can be partially absorbed, reflected towards the outer space or diffused in all directions when it encounters gas molecules (like oxygen, carbon-dioxide, ozone), vapours (water) and atmospheric particles. The portion of the solar radiation hitting the Earth's surface after all the absorption, reflection and diffusion effects is the *direct radiation*. The portion of the total solar radiation scattered in the atmosphere but still reaches the surface is the *diffuse radiation*.

The solar radiation spectrum (Figure 4.1) displays the different wavelengths present in the solar spectrum as well as the absorption phenomena caused by the presence of water and carbon-dioxide in the atmosphere.

In order to properly consider the effect caused by the thickness of the atmosphere through which the solar radiation travels, the *Air Mass index* is defined as:

$$AM = \frac{P}{P_0 \sin\theta} \tag{4.2}$$

where P is the atmospheric pressure, P_0 is the reference pressure of 0.1013 MPa and θ is the Sun's elevation angle with respect to the horizon.

AM is equal to 1 when the Sun is perpendicular to the surface at sea level ($P = P_0$, $\theta = 90°$). AM at sea level is equal to 2 when the Sun forms an angle of 30 ° with the horizon. At European latitudes, AM is set at 1.5, a value used also as a reference for the tests conducted in laboratories on photovoltaic cells. AM equals to 0 when the radiation is measured on the outer surface of the atmosphere with no attenuation whatsoever.

Solar radiation reaches its maximum in the visible range of 0.48 μm and decreases rapidly in the ultraviolet range. The descent towards the infrared frequencies is slower. The radiation wavelengths reaching the terrestrial surface mainly fall in the range of 0.2–2.5 μm and they are distributed according to the following:

- 0.2–0.38 μm, ultraviolet range (containing 6.4% of the total energy shown in the spectrum);
- 0.38–0.78 μm, visible range (carrying 48% of the total energy);
- 0.78–10 μm, infrared range (with 45.6% of the energy mentioned above).

The electromagnetic energy of the Sun is constituted by photons which are characterised by a *wave-particle duality* behaviour. Normally, the photons hitting a material transfer part of their energy to the material particles whose internal energy is increased as a consequence. For some materials, apart from this increase in the internal energy, other phenomena can also be observed, such as:

- *photoexcitation*: the shift of an electric charge to a higher energy state in an atom or a molecule;
- *photoionization*: the ionization of an atom or a molecule;
- *photoelectric effect*: the release of electrically charged particles from a material;
- *photovoltaic effect*: the creation of an electric field between two different materials (usually semiconductors).

4.2 Photovoltaic Effect, Semiconductors and the p-n Junction

Solar irradiance is a renewable energy source that fits the Moriarty-Honnery criteria described in Section 1.3. As for the conversion from electromagnetic to electric energy, semiconductors are the most widely-adopted material to perform such procedure, with silicon being the most popular element utilised among many others.

In the Earth's crust, silicon is the second most abundant element after oxygen, constituting 25.7% of the crust by mass. It can be found in clay, feldspar, granite, quartz and sand, mostly in the forms of silicon dioxide, silicates and aluminium silicates.

Silicon has atomic number 14 and possesses 4 electrons in the valence band. In the crystal structure of pure silicon, every atom is covalently bonded to other four atoms in a stable crystalline configuration. Two atoms next to a pure silicon crystal structure share a couple of electrons in common. The covalent bond is the strongest chemical bond of all and can be broken only by an exchange of energy larger than the *energy gap* (E_g), so that the electrons in the valence band can move to the conduction band. For a silicon atom this energy equals 1.12 eV (1 eV = $1,602 \times 10^{-19}$ J), an intermediate value between the amounts of energy required for an insulator and a conductor.

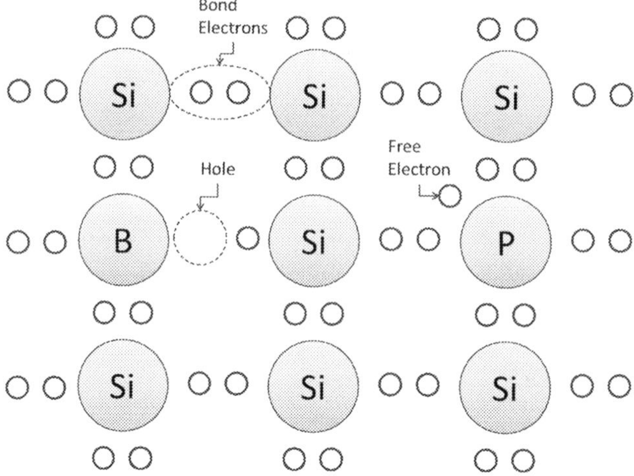

Fig. 4.2. The crystalline lattice of *n-type* and *p-type* doped silicon

If the electron is able to absorb such energy and move to the valence band, it leaves a hole behind which can be easily filled up by another neighbouring electron. The filling electron in turn leaves another hole behind. With this mechanism, the hole behaves in a way similar to a mobile particle with a positive charge, mirroring the movement of the electrons who generate them.

It is possible to modify the electric properties of silicon by introducing impurities into the element crystal structure. This procedure of contamination of the intrinsic structure of silicon is called *doping*.

As Figure 4.2 demonstrates, the doping elements to be introduced into silicon belonging to Group III or Group V in the periodic table.

A *p-type semiconductor* can be obtained by adding small quantities of boron (Group III) in the silicon lattice. The result is that one electron will be missing from one of the four covalent bonds, creating a *hole* in the structure. Boron is thus defined as the *acceptor* material.

An *n-type semiconductor* instead can be created by injecting phosphorus (Group V) in the silicon lattice. The result is that one phosphorous electron will not participate in the covalent bonds and becomes free to move in the crystal three-dimensional structure. This electron will be weakly bonded to the silicon lattice and can be shifted to the conduction band with a much lower energy compared to other covalently-bonded electrons. Phosphorus therefore becomes the *donor* material.

Since the quantity of the dopants is very low, they are considered to be "impurities" that modify the electric, but not the chemical, behaviour of silicon. The dopant quantity is measured in atoms/cm^3 and varies from 10^{13} to 10^{20} atoms/cm^3 with respect to a density of 10^{22} atoms/cm^3 of the silicon atoms in the lattice structure.

When a p-type and an n-type semiconductor are joined together, a *p-n junction* is created. The concentration gradient of the electrons and the holes causes the electrons to diffuse from the n-type to the p-type semiconductor while the holes distribute in

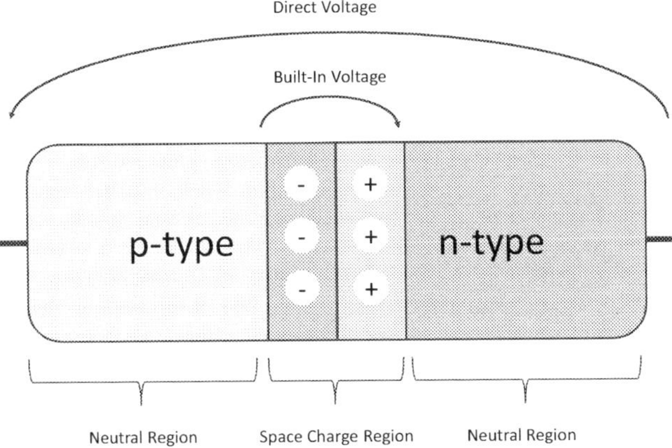

Fig. 4.3. Schematic of the p-n junction

the opposite direction. When the electrons from the n-type region enter the p-type region, they recombine with the holes and vice-versa. This creates a *depletion layer* (or *space-charge region*) on the interface of the two semiconductors where an electric potential is built up because the phosphorus atoms are left with positive charges and boron atoms with negative ones. The *built-in voltage* thus created at the junction becomes opposed to the diffusion currents which contribute to establish the depletion region (Figure 4.3).

The p-n junction constitutes a particular electric circuit called *diode*. The p-type semiconductor is the anode while the n-type is the cathode. The built-in voltage in the depletion region of a silicon semiconductor is between 0.6 to 2 V while for a germanium semiconductor it is in the range of 0.3 to 0.8 V.

If the diode is connected to an external electric circuit, it becomes conductive if the anode voltage is positive compared to the cathode (*direct polarization*) and the applied voltage is higher than the built-in voltage. Instead, when the anode voltage is negative (*inverse polarization*), the diode becomes practically non conductive. If such inverse voltage is higher than that of the diode *breakdown voltage*, the diode becomes conductive at the expense of potential damages to the junction[2].

The electric current in a diode is expressed analytically as:

$$I = I_c + I_d \qquad (4.3)$$

where:

$$I_c = qA\,\mu_n\,n\varepsilon + qA\,\mu_p\,p\varepsilon \qquad (4.4)$$

and:

$$I_d = qA\,D_n\,\frac{dn}{dx} - qA\,D_p\,\frac{dp}{dx}. \qquad (4.5)$$

---

[2] A *Zener diode* is a special diode designed to conduct currents in reverse. It is widely employed, for example, as a voltage stabilizer in electric operations where a precision voltage reference is required.

I_c is the current from the applied electric field, I_d is the diffusion current, A is the section of the semiconductor, q is the charge of an electron[3], n is the concentration of the electrons, p is the concentration of the holes, ε is the applied electric field, μ_n is the mobility of the electrons, μ_p is the mobility of the holes, D_n is the diffusion constant of the electrons and D_p is the diffusion constant of the holes.

4.3 Crystalline Silicon Photovoltaic Cells

After many theoretical elaborations and experiments, the first commercial photovoltaic cell was produced in 1954 at Bell laboratories from a junction on monocrystalline silicon, which became the foundation for the development of the current photovoltaic cells based on semiconductors. At that time, because of the high production costs, the first successful applications were mainly limited to military and aero-spatial fields. With the birth of new production technologies in the years to follow, the modules were gradually introduced into a wider consumer market.

When a semiconductor material is exposed to sunlight, the interaction between the photons and the electrons in the valence band enables some electrons to pass to the conduction band. This phenomenon contributes to the formation of electron–hole pairs inside the semiconductor and creates a voltage differential on the electrodes connected to the junction. A semiconductor photovoltaic cell is therefore a p-n junction diode which generates electron-hole pairs when exposed to a source of light.

When a valence electron interacts with a sufficiently energetic photon, it increases its energy and jumps to the conduction band. The electrons diffuse towards the n-doped semiconductor and the holes towards the p-doped semiconductor due to the diode built-in voltage. If the two electrodes are connected by an electric wire, the equilibrium is established by the flow of electrons from the n-doped to the p-doped semiconductor. This is how the interaction with light generates a continuous flow of electric current.

The *I-U* relation of the diode in direct conduction is expressed as:

$$I_D = I_0 \left[\exp\left(\frac{qU}{NKT} \right) - 1 \right] \tag{4.6}$$

where q is the charge of an electron, K is the Boltzmann constant [4], T is the temperature, I_0 is the inverse saturation current in the diode and N is a coefficient between 1 and 2, depending on the phenomena of generation and recombination occurring in the depletion layer with N = 1 for an ideal diode. $U_T = KT/q$ is the so-called *thermal voltage*.

The analytical expression of I_0 is:

$$I_0 = A_0 T^3 \exp\left(\frac{-E_g}{KT} \right) \tag{4.7}$$

where A_0 is a constant that depends on the type of semiconductor chosen.

---

[3] q = 1,602 × 10^{-19} C.
[4] K = 1.38 × 10 − 23 J/K.

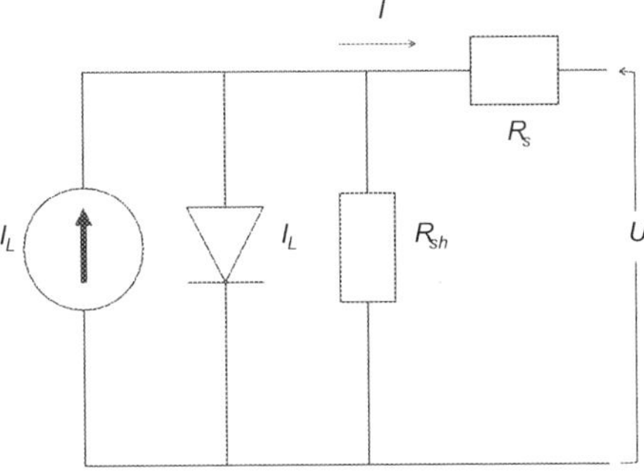

Fig. 4.4. The equivalent circuit of the single-diode model in the photovoltaic cell

In order to analyse the behaviour of the p-n junction under solar radiation, the *single-diode model* is used (Figure 4.4).

From the Kirchhoff's laws, the equation of the cell subject to incident light becomes:

$$I = I_L - I_D - I_{Rsh} = I_L - I_0 \left[\exp\left(\frac{U + IR_s}{a} \right) - 1 \right] - \frac{U + IR_s}{R_{sh}} \tag{4.8}$$

in which I_L is proportional to the number of the photons that activate the photovoltaic effect and $a = NKT/q$. R_s is the parasitic *series resistance* caused by the internal resistance of the cell and the ohmic resistance from the electric contacts and their interface with the silicon. At high irradiation levels, the voltage drop caused by the series resistance can seriously affect the power yield. The other parasitic *shunt resistance* R_{sh} represents the power loss caused by the electric current diversion inside the cell caused and such diversion is attributed to the cell manufacturing defects and irregularities. The shunt resistance reduces the voltage output of the solar cell, particularly at low irradiation levels.

Figure 4.5 shows the *I-U* characteristic curve of a photovoltaic cell. The voltage U reaches a maximum value of U_{oc} when in open circuit, while the current I in short-circuit conditions reaches the value I_{sc}.

In short-circuit conditions the voltage is zero and $I = I_{sc}$, for this reason the characteristic equation becomes:

$$I_{sc} = I_L - I_0 \left[\exp\left(\frac{I_{sc}R_s}{a} \right) - 1 \right] - \frac{I_{sc}R_s}{R_{sh}}. \tag{4.9}$$

In open-circuit conditions $I_{sc} = 0$ and $U = U_{oc}$, so the characteristic equation becomes:

$$0 = I_L - I_0 \left[\exp\left(\frac{U_{oc}}{a} \right) - 1 \right] - \frac{U_{oc}}{R_{sh}}. \tag{4.10}$$

At the *maximum power point* (MPP), which corresponds to the coordinates (I_{MP}, U_{MP}), the equation becomes:

$$I_{MP} = I_L - I_0 \left(\exp\frac{U_{MP} + I_{MP}R_s}{a} - 1 \right) - \frac{U_{MP} + I_{MP}R_s}{R_{sh}}. \tag{4.11}$$

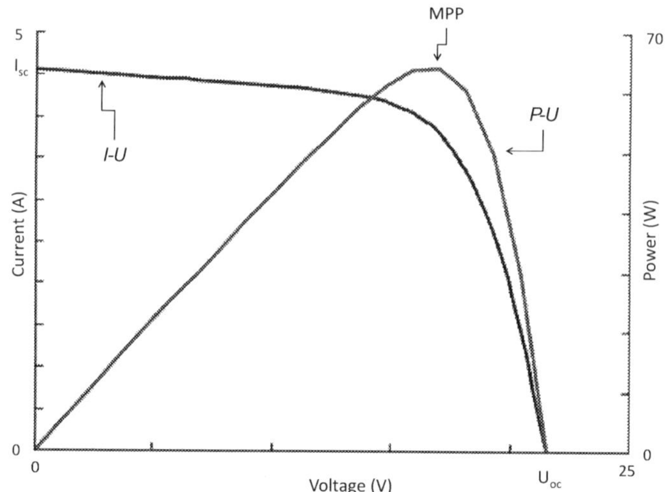

Fig. 4.5. The *I-U* and *P-U* characteristic curves of a photovoltaic cell

At the maximum power point the derivative of P is zero with respect to U (Figure 4.5):

$$\left.\frac{dP}{dU}\right|_{P=P_{MPP}} = 0 = I_{MP} + U_{MP}\left.\frac{dI}{dU}\right|_{I=I_{MP}} \tag{4.12}$$

hence:

$$\left.\frac{dI}{dU}\right|_{I=I_{MP}} = \frac{-\frac{I_0}{a}\exp\frac{U_{MP}+I_{MP}R_s}{a} - \frac{1}{R_{sh}}}{1 + \frac{I_0 R_s}{a}\exp\frac{U_{MP}+I_{MP}R_s}{a} + \frac{R_s}{R_{sh}}}. \tag{4.13}$$

The *Filling Factor* estimates the construction quality of the cell:

$$FF = \frac{I_{MP}U_{MP}}{I_{sc}U_{oc}}. \tag{4.14}$$

The closer to 1 the filling factor is, the better the quality of the cell.

Another key criterion to evaluate the cell performance is the conversion efficiency η, defined as the ratio between the maximum power generated by the cell and the incident solar radiation per unit area G_T multiplied by the active area of the cell A:

$$\eta = \frac{I_{MP}U_{MP}}{G_T A}. \tag{4.15}$$

Multiplying the above numerator and the denominator by $(q U_{oc} I_{sc} E_g)$ it becomes:

$$\eta = \frac{I_{MP}U_{MP}}{G_T A}\frac{q U_{oc} I_{sc} E_g}{q U_{oc} I_{sc} E_g} = \frac{I_{sc}E_g}{qA G_T}\left(\frac{I_{MP}U_{MP}}{I_{sc}U_{oc}}\right)\frac{q U_{oc}}{E_g} \tag{4.16}$$

which shows how the efficiency depends also on the filling factor. High quality cells, with a value of the filling factor close to 1, will increase conversion efficiency.

4.4 Other Cell Technologies

The single-diode model described above is only applicable to crystalline silicon cells. Other types of technologies are modelled differently according to their respective semiconductor structures and physical properties. For example, they can be based on amorphous silicon (a:Si), which is doped in tandem or triple semiconductor junctions, or by elements like cadmium, tellurium, copper, indium, gallium and selenium. By combining junctions based on such elements, CdTe, CIGS or CIS cells can be obtained. The semiconductor layers are grown as *thin-films* one over the other with a non-crystalline structure.

Amorphous silicon junctions differ from other technologies by the existence of an *intrinsic* layer (p-i-n junction), therefore a different model applies for an a:Si film cell [9]. Since recombination losses occur in the i-layer of the p-i-n junction, a term is added to the single diode model representing a current leak, which is a function of the photocurrent and the cell voltage.

$$I_{rec} = I_L - \frac{d_i^2}{\mu_{eff}\left(U_{bi} - U - IR_s\right)} \tag{4.17}$$

where d_i is the thickness of the i-layer, μ_{eff} is the diffusion length of the charge carriers and U_{bi} the built in voltage of the junction (0.9 V for a:Si).

4.5 Conversion Losses

The photovoltaic cell manages to convert in electricity only a small portion of the total incident solar radiation. For instance, a polycrystalline silicon photovoltaic cell has a conversion efficiency around 14% and a monocrystalline cell delivers a slightly higher rate over 17%. Thin film cells instead show an efficiency around 10-11% while other new technologies seem to be able to reach over 25-30%.

A part of the incident solar radiation is reflected away by the metal front grid electric contacts or because of the overly small angle of incidence of the sunlight. The reflected light therefore cannot pass through the glass to interact with the electrons in the junction.

The energy of the incident photons varies according to their frequencies and therefore not all of them can successfully bring the electrons to the conduction band. The maximum wavelength λ_{max} above which the photons do not have sufficient energy to activate the photovoltaic effect is given by:

$$\lambda_{max} = \frac{hc}{E_g} \tag{4.18}$$

where E_g is the energy gap, h is the Planck constant and c is the speed of light. Since λ_{max} equals 1.11 μm for silicon, 22% of the energy contained in the spectrum is inefficient to generate electron-hole pairs, as shown in Table 4.1. These photons dissipate the excess energy in the form of heat which increases the cell temperature and the ohmic resistance.

Table 4.1. Spectral distribution of solar energy

Wavelength (μm)	Percentage
$0.3 < \lambda < 0.5$	17%
$0.5 < \lambda < 0.7$	28%
$0.7 < \lambda < 0.9$	20%
$0.9 < \lambda < 1.1$	13%
$\lambda > 1.1$	22%

On the contrary, photons that are too energetic can pass through the cell structure without interacting with the electrons in the junction. The fraction of the absorbed incident energy can be calculated by:

$$\frac{E_{absorbed}}{E_{incident}} = 1 - \exp(-\alpha d) \qquad (4.19)$$

where d is the distance that the radiation travels in the crystal structure before being completely absorbed. α assumes a finite value which depends on the wavelengths of the incident solar radiation. The lower α becomes, the farther away from the surface the photons are absorbed.

Not all the electron-hole pairs generated are able to diffuse towards the electrodes due to the recombinations that occur before they reach the electrodes.

The depth of the junction plays an important role in the design of photovoltaic cells. The *diffusion length* of an electron is the average distance that an electron travels from the moment it passes to the conduction band to when eventually it recombines. Depending on the evaluation or the measure of the diffusion length, the depth of the junction is reduced accordingly. Consequently there is a trade-off between keeping the junction large enough to facilitate the absorption of high frequency photons and reducing the junction to prevent recombinations. With all the above elements considered, the depth of the junction is usually maintained at around 0.2–0.4 μm.

Finally, as already mentioned, another type of energy loss is caused by the parasitic resistances.

4.6 Changes in the *I-U* Curve

Changes in the incident solar irradiance and in the temperature of the photovoltaic cell modify the cell *I-U* curve. If the solar irradiance increases, the short-circuit current I_{sc} augments while there are no noticeable changes in the open-circuit voltage (Figure 4.6). Therefore, the power converted by the cell increases with the rise in solar irradiance.

An increase in the cell temperature causes a reduction of the open-circuit voltage U_{oc} with small changes in the short-circuit current I_{sc} (Figure 4.7). Therefore, the power converted by the cell decreases when the cell temperature rises.

Accordingly, the favourable weather conditions are those capable of delivering a high solar irradiance while helping maintain a low cell operating temperature.

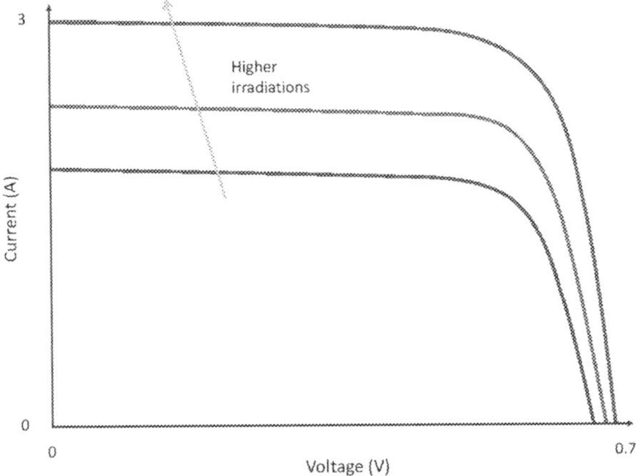

Fig. 4.6. The variations of the *I-U* curve according to the degree of radiation (in W/m^2) at 40 °C

4.7 Photovoltaic Cells and Modules

The basic component of a semiconductor photovoltaic system is the cell. For crystalline silicon the cell is usually square-shaped with a surface of around 100 cm^2 and produces an electric current around 3–7 A with a voltage of 0.5 V. The power output is 1.5 W in standard test conditions.

In order to obtain the desired voltage and current output level, the cells are connected in series and in parallel. The assembly becomes the *photovoltaic module*. To protect the cells while at the same time allowing the sunlight to pass through, the

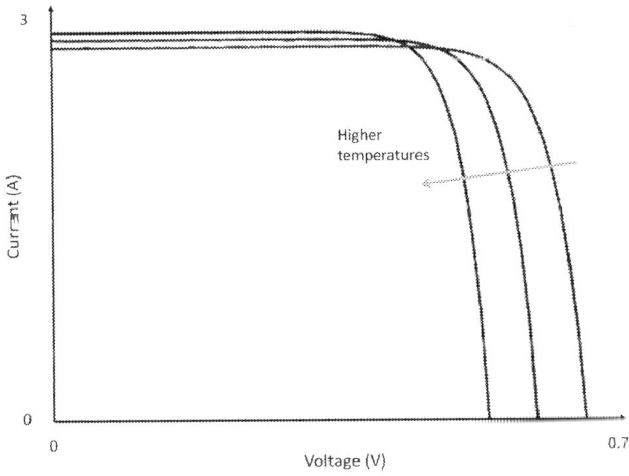

Fig. 4.7. *I-U* curves with varying temperatures

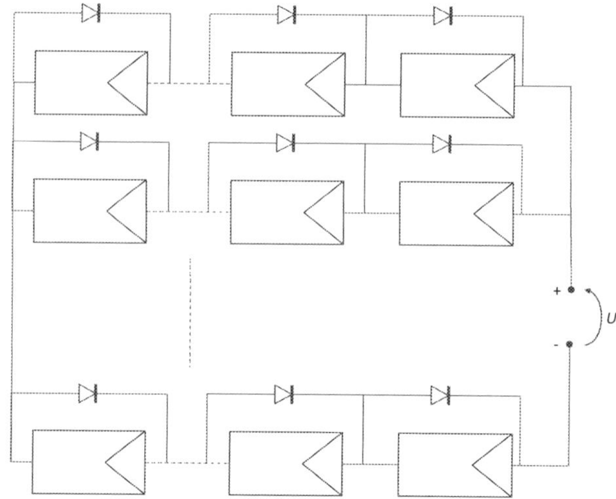

Fig. 4.8. The connection layout of the modules in a photovoltaic system

modules are covered at the front with a layer of tempered glass as thick as 4 mm . The mechanical properties of the covering glass need to ensure its resistance to high-impact collisions from atmospheric agents like hailstones. To improve light transmission, the glass is specially manufactured with low iron content and titanium oxide coating.

A thin layer of transparent ethylene-vinyl acetate (EVA) is placed between the glass and the photovoltaic cells to avoid direct contact of the two and to eliminate any interstices. The back of the module is laminated with another layer of EVA and wrapped with a Tedlar sheet. The assembly is usually framed with an aluminium profile.

The cells are electrically connected by metal finger bars, while the outer connections are routed to a connection box hosting by-pass diodes for cell protection.

Within a module, the cells are connected together with by-pass diodes to prevent the current from flowing in cells that are not converting electric current. This could happen with a partial shadowing on the module surface or with damages to the cells themselves. In this case, the electricity produced from other cells would flow through non-productive cells and heat them up because of the Joule effect with potential damages to the module.

The modules can be connected in series or in parallel to provide the desired power output of the whole photovoltaic power plant. Figure 4.8 shows the connection layout of the modules in a photovoltaic plant. The bypass diodes can be seen as connected in parallel to the modules. An electric protection (i.e. a fuse) is installed at the start of each string.

Since the energy produced by a photovoltaic module depends mostly on its temperature and the incident solar radiation, in order to compare the performances among different cells and modules a special measuring unit, the *Watt-peak* (W_p), is adopted.

The Watt-peak indicates the power delivered by the module at the *Standard Test Conditions* (STC), namely:

- cell temperature = 25 °C;
- solar radiation intensity = 1 kW/m^2;
- AM = 1.5.

If modules with different characteristics are connected in series, the output of the whole system is conditioned by the modules with lower performance. This problem is known as *mismatch* and can usually be avoided by connecting together the modules with similar electric characteristics. Such characteristics vary from module to module and can be obtained by performing tests (*flash-tests*) in standard conditions by measuring the *I-U* characteristic curve and the main electric parameters.

The only way to prevent any voltage output from the photovoltaic module is to cover its surface completely. This can raise safety concerns when, for example, the voltage supply from a photovoltaic plant needs to be shut down because of the break-out of a fire or other types of emergencies. Special solutions need to be taken into account to reduce such risk.

4.8 Types of Photovoltaic Plants

A photovoltaic system consists of an ensemble of photovoltaic generators, control systems and power conditioners. Depending on the type of connection to the grid, these systems are distinguished between *stand-alone* and *grid-connected* plants.

Stand-alone systems supply electricity to electric loads isolated from the grid. Usually they are used to power mountain chalets, telecommunication stations, street lighting, water pumping stations in developing countries or wherever the grid cannot reach the loads. In order to guarantee the continuity of electricity supply, it is necessary to adopt energy storage systems such as batteries, electrochemical accumulators and relative charge regulators.

The grid-connected photovoltaic systems instead provide electricity directly to the grid. This type of system does not need energy storage since the continuity of the supply is guaranteed by the grid itself. It is therefore not obligatory to size the plant according to load needs and, in this way, the design of the plant structure can be more flexible and enjoy more architectural freedom.

Photovoltaic plants can be in various sizes from small residential plants of a few kW$_p$ to larger installations in the range of multi MW$_p$. Figure 4.9 displays a large-scale ground plant installed on fixed structures.

The panels of these photovoltaic systems can be either *fixed* or equipped with *tracking* technologies. Fixed panels are installed upon a static support. If the plant is equipped with tracking devices, the module surface is able to follow the position of the Sun during the course of the day in order to maximize the quantity of the incident radiation energy the panels receive.

The electric architecture of a generic photovoltaic system is shown in Figure 4.10. The module strings are connected to the inverter usually with a protection instrument such as a blocking diode, a fuse or a circuit breaker, so that no inverse current will pass

Fig. 4.9. Ground-based photovoltaic plant [10]

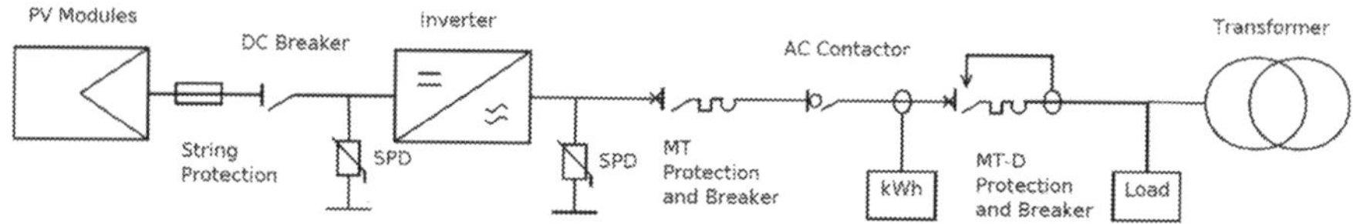

Fig. 4.10. Architecture of a photovoltaic field, DC and AC side (Reproduced with permission from [12])

through the string when the inverter is impeded to function properly as a generator. A protection is also needed to prevent the cable from being damaged or burnt when its current capacity is exceeded when short-circuits occur. After the string protections, DC circuit breakers give the user the possibility to disconnect the photovoltaic field even under solar irradiation, for maintenance or safety reasons. The inverter converts the DC to AC necessary to supply the loads or energy to the electricity distribution grid. Surge protection devices (SPD) shield the inverter from voltage surges resulted from lightnings. After the inverter, on the AC side, a series of circuit breakers protect the AC lines as per the normal electric design practice. A metering unit is installed to measure the energy converted by the photovoltaic plant. Finally, a transformer is employed in case a voltage increase is needed to inject energy in the medium-voltage or high-voltage grid.

In this electric circuit, the conversion operations of the inverter and the transformers add to the total power loss of the system. However, the amount of power lost in this aspect is relatively low since these two devices still possess rather high conversion efficiencies (around 94–98% for the inverter and 97–98% for the transformer).

4.9 Radiation on the Receiving Surface

To increase energy conversion, the module surface must be positioned in a way to intercept the highest possible amount of solar radiation. The following angles and formulas are essential for computing the solar radiation captured by the total photovoltaic surface.

The Sun's *elevation h* (and its complementary angle *z, zenith*) is the angle seen by the observer between the position of the Sun and the horizontal plane (for *z*, the perpendicular to the horizontal plane).

The *azimuth* α, (or *orientation*) is the angle of the Sun around the horizon on the horizontal plane. North is usually set at an azimuth of $0°$, while South is $180°$.

The *tilt* β (or *inclination*) is the angle formed by the plane of the receiving surface with respect to the horizontal plane.

The *latitude* ϕ is the latitude of the site in question.

The *declination* δ is the angle that the Sun's rays form with the equatorial plane, which is positive when the Sun stays above the plane and negative in the opposite case. It is calculated at a daily rate with the following:

$$\delta = 23.44° \times \sin\left[\frac{360°\,(284+n)}{365}\right] \tag{4.20}$$

where n is the number of the day of the year.

The *solar hour angle* ω is the angular distance between the Sun's position at noon and the other different positions when the Sun shifts along the trajectory on the celestial vault. Such value is positive before noon since the earth rotates at $15°$ per where t stands for the hour and:

$$\omega_s = \cos^{-1}(-\tan\phi\,\tan\delta) \tag{4.21}$$

is the hour angle at dawn.

The total solar irradiance G_T (in W/m^2) that the photovoltaic surface receives is calculated with:

$$G_T = G_b R_b + G_d R_d + G_g R_r \tag{4.22}$$

where G_b is the direct solar irradiance on the horizontal plane, G_d is the diffuse solar irradiance on the plane and G_g is the reflected solar irradiance, equal to:

$$G_g = G_b + G_d \tag{4.23}$$

and R_d and R_r are the factors that account for the portions of the diffuse and the reflected solar irradiance received by the module surface, given by:

$$R_d = \frac{1+\cos\beta}{2} \qquad R_r = \rho\,\frac{1-\cos\beta}{2}. \tag{4.24}$$

Table 4.2. Albedo values of some different ground surfaces

Type of land surface	Albedo
Snow	0.75
Grass	0.29
Soil	0.14
Tar	0.13

ρ is the surface reflectance coefficient (or *albedo*) that depends on the type of ground surface on which the photovoltaic modules are installed (Table 4.2).

The term R_b is the tilt factor for the direct solar radiation. When the light-receiving plane is facing true south ($\alpha = 180°$), R_b becomes[5]:

$$R_b = \frac{\sin\delta\ \sin(\phi - \beta) + \cos\delta\ \cos\omega\ \cos(\phi - \beta)}{\sin\phi\ \sin\delta - \cos\phi\ \cos\delta\ \cos\omega}. \tag{4.25}$$

In order to compute the solar radiation on the photovoltaic modules, it is necessary to know the direct and the diffuse radiation on the horizontal plan at an hourly rate. These terms can be obtained with the following equations:

$$G_b = \frac{180°}{24} H_{bo} \frac{\sin\delta\ \sin\phi + \cos\delta\ \cos\omega\ \cos\phi}{\omega_s\ \sin\delta\ \sin\phi - \cos\delta\ \cos\omega_s\ \cos\phi} \tag{4.26}$$

$$G_d = \frac{180°}{24} H_{do} \frac{\cos\omega - \cos\omega_s}{\sin\omega_s - \omega_s\ \cos\omega_s} \tag{4.27}$$

where H_{bo} is the monthly average daily direct solar radiation on the horizontal plan (in kWh/m^2) and H_{do} is the monthly average daily diffuse radiation.

Solar radiation also depends on the geographical coordinates of the site. Since these are average data valid for the whole month, for calculations it is simpler to adopt a *significant day of the month*, which is defined as the day best representing the weather and solar characteristics of the month.

4.10 Determination of the Operating Point

As previously mentioned, it is possible to define the behaviour of a photovoltaic cell using the *single-diode* model with the following characteristic equation:

$$I = I_L - I_D - I_{Rsh} = I_L - I_0 \left[\exp\left(\frac{U + IR_s}{a}\right) - 1 \right] - \frac{U + IR_s}{R_{sh}} \tag{4.28}$$

where the five parameters I_L, I_0, a, R_s and R_{sh} depend on the technology, the temperature of the cell and the incident solar radiation.

---

[5] For every value of α, the numerator of the equation becomes: ($\sin\delta\ \sin\Phi\ \cos\beta - \sin\delta\ \cos\Phi\ \sin\beta\ \cos\alpha + \cos\delta\ \cos\Phi\ \cos\beta\ \cos\omega + \cos\delta\ \sin\Phi\ \sin\beta\ \cos\alpha\ \cos\omega + \cos\delta\ \sin\beta\ \sin\alpha\ \sin\omega$).

These five parameters can be empirically evaluated using the data from the characteristic curves. These data are retrieved by flash-tests that in standard conditions (subscript STC) provide:

- the short circuit current $I_{sc,STC}$;
- the open circuit voltage $U_{oc,STC}$;
- the current and the voltage at maximum power output $I_{MP,STC}$, $U_{MP,STC}$;
- the current $I_{x,STC}$ for a voltage $U_{x,STC} = 0.5\,U_{oc,STC}$;
- the current $I_{xx,STC}$ for a voltage $U_{xx,STC} = 0.5\,(U_{oc,STC} + U_{MP,STC})$.

In order to evaluate the five parameters in standard conditions, the following system of five non-linear equations needs to be structured for the five unknowns $I_{L,STC}$, $I_{0,STC}$, a_{STC}, $R_{s,STC}$ and $R_{sh,STC}$ to be solved numerically:

$$
\begin{cases}
I_{sc,STC} = I_{L,STC} - I_{0,STC}\left[\exp\left(\dfrac{I_{sc,STC}\,R_{s,STC}}{a_{STC}}\right) - 1\right] - \dfrac{I_{sc,STC}\,R_{s,STC}}{R_{sh,STC}} \\[4mm]
0 = I_{L,STC} - I_{0,STC}\left[\exp\left(\dfrac{U_{oc,STC}}{a_{STC}}\right) - 1\right] - \dfrac{U_{oc,STC}}{R_{sh,STC}} \\[4mm]
I_{MP,STC} = I_{L,STC} - I_{0,STC}\left[\exp\left(\dfrac{U_{MP,STC}+I_{MP,STC}\,R_{s,STC}}{a_{STC}}\right) - 1\right] - \dfrac{U_{MP,STC}+I_{MP,STC}\,R_{s,STC}}{R_{sh,STC}} \\[4mm]
I_{x,STC} = I_{L,STC} - I_{0,STC}\left[\exp\left(\dfrac{U_{x,STC}+I_{x,STC}\,R_{s,STC}}{a_{STC}}\right) - 1\right] - \dfrac{U_{x,STC}+I_{x,STC}\,R_{s,STC}}{R_{sh,STC}} \\[4mm]
I_{xx,STC} = I_{L,STC} - I_{0,STC}\left[\exp\left(\dfrac{U_{xx,STC}+I_{xx,STC}\,R_{s,STC}}{a_{STC}}\right) - 1\right] - \dfrac{U_{xx,STC}+I_{xx,STC}\,R_{s,STC}}{R_{sh,STC}} .
\end{cases}
\tag{4.29}
$$

Since it is unlikely for the cell to work in STC, it is essential to determine these parameters in all possible temperature and irradiance conditions. For this purpose the following equations can be used, the first being:

$$
I_L = \frac{G_T}{G_{T,STC}}\left[I_{L,STC} + \mu_{I_{sc}}\left(T_p - T_{p,STC}\right)\right]
\tag{4.30}
$$

where G_T is the total solar irradiance in W/m^2, $G_{T,STC}$ is equal to 1000 W/m^2, $\mu_{I_{sc}}$ is the gradient of the short circuit current with respect to the temperature, T_p is the temperature of the cells and $T_{p,STC}$ is the temperature of the cells in standard conditions (25 °C). The second equations is.

$$
I_0 = I_{0,STC}\left(\frac{T_p}{T_{p,STC}}\right)^3 \exp\left[\frac{E_g N_s}{a_{STC}}\left(1 - \frac{T_{p,STC}}{T_p}\right)\right]
\tag{4.31}
$$

where E_g is the energy gap of the materials of the cells and N_s is the number of the cells connected in series.

The remaining parameters are given by:

$$
a = a_{STC}\,\frac{T_p}{T_{p,STC}}
\tag{4.32}
$$

$$R_s = R_{s,STC} \tag{4.33}$$

$$R_{sh} = R_{sh,STC}\, \frac{G_{T,STC}}{G_T}. \tag{4.34}$$

In a photovoltaic system, a specially designed algorithm (*Maximum Power Point Tracker* – MPPT) always guarantees that the operating point is the couple of values *I-U* that correspond to the delivery or the maximum available power. This can be obtained by numerically solving the following system:

$$\begin{cases} I_{MP} = I_L - I_0 \left[\exp\left(\frac{U_{MP}+I_{MP}R_s}{a}\right) - 1\right] - \frac{U_{MP}+I_{MP}R_s}{R_{sh}} \\[4mm] \left.\frac{dP}{dV}\right|_{P=P_{MPP}} = 0 = I_{MP} + U_{MP}\, \dfrac{-\frac{I_0}{a}\exp\left(\frac{U_{MP}+I_{MP}R_s}{a}\right) - \frac{1}{R_{sh}}}{1 + \frac{I_0 R_s}{a}\exp\left(\frac{U_{MP}+I_{MP}R_s}{a}\right) + \frac{R_s}{R_{sh}}} \end{cases}. \tag{4.35}$$

The temperature T_p of the cells of the panel can be measured by a thermal balance or by the following experimentally-established formula:

$$T_p = T_{p,b} + \frac{G_T}{G_{NOCT}}\, \Delta T_{p,NOCT} \tag{4.36}$$

where $\Delta T_{p,NOCT}$ is the temperature difference between the cells and the back of the panel in standard conditions, G_{NOCT} is the intensity of the total incident radiation in *Nominal Operating Cell Temperature* conditions ($G_{NOCT} = 800$ W/m2 and $T_a = 20°C$). $T_{p,b}$ is the back surface temperature of the module given by a linear regression as:

$$T_{p,b} = T_a + \exp\left(a_c + b_c\, W_s\right) \tag{4.37}$$

where T_a stands for the ambient temperature, a_c and b_c are two empirical constants of the module and W_s is the wind velocity.

To obtain the ambient temperature the following hourly-rate formula can be used:

$$T_a = 0.5 \left[\left(T_{a,max} + T_{a,min}\right) + \left(T_{a,max} - T_{a,min}\right)\sin\frac{2\pi\left(t - t_p\right)}{24}\right] \tag{4.38}$$

where t is the hour of the day in which the temperature is measured and t_p is the difference in hours between when the temperature reaches the maximum during the day ($T_{a,max}$) and when the temperature lowers to the minimum ($T_{a,min}$). Similar to the case of evaluating the probability of solar availability, it is easier to refer to the monthly average temperature for $T_{a,max}$ and $T_{a,min}$. These monthly average numbers are available from meteorological databases while t_p can be found on specialized weather websites month by month.

With this method it is possible to compute the characteristic parameters of the strings of the modules and then obtain the operating points of the entire photovoltaic system.

References

1. Cooper P I (1969) The absorption of solar radiation on solar stills. Solar Energy 12 (3):333–346
2. Duffie J D, Beckman W A (1980) Solar Engineering of Thermal Processes. Wiley, New York
3. Garg H P (1982) Treatise on Solar Energy. Wiley, New York
4. Green M A (1998) Solar cells: Operating Principles, Technology and System Applications. University of New South Wales
5. King D L (2004) Photovoltaic Array Performance Model. Sandia National Laboratories in Albuquerque, New Mexico, Report 87185–0752
6. Lasnier F, Ang T G (1990) Photovoltaic Engineering Handbook. Hilger, Bristol
7. Liu B Y H, Jordan R C (1963) The long term average performance of flat plate solar energy collectors. Solar Energy 7:53–70
8. Luque A, Hegedus S Eds (2003) Handbook of Photovoltaic Science and Engineering. John Wiley & Sons Ltd, Chichester
9. Mertens J, Asensi J M, Voz C, Shah A V et al (1998) Improved equivalent circuit and analytical model for amorphous silicon solar cells and modules. IEEE Transactions on Electron Devices 45(2):423-429
10. OhWeh (photographer) [CC-BY-SA-2.5 (www.creativecommons.org/licenses/by-sa/2.5)], via Wikimedia Commons, http://commons.wikimedia.org/wiki/File:SolarparkTh%C3%BCngen-020.jpg
11. Scheer H (1999) Solare Weltwirtschaft. Verlag Antje Kunstmann GmbH, München
12. Zini G, Mangeant C, Merten J (2011) Reliability of large-scale grid-connected photovoltaic systems. Renewable Energy 36 (9):2334–2340

5

Wind Energy

Together with solar radiation, wind power represents an important source for the hybrid renewable energy systems. Wind aerogenerators convert the kinetic energy in the wind to electric energy either for immediate grid connection or for storage. However, wind energy is also highly volatile, therefore a sound storage solution is required in order to reliably supply the load.

5.1 Introduction

The radiation from the Sun heats the Earth's atmosphere differently around the globe. The thermal gradient between the air at the equator and the poles causes a differential in the atmospheric air pressure that generates the flow of large masses of air, the *wind*.

The kinetic energy in the wind can be intercepted and converted into electric energy with wind turbines or *aerogenerators* (Figure 5.1) that generally consist of the following components:

- tower;
- rotor;
- nacelle;
- electric generator;
- brake system;
- rotation adapter;
- control system.

The tower is usually constructed in a tubular or lattice structure. It supports the nacelle and the rotor blades at the desired operation height. Its foundation and fixture structures vary depending on the type of land surface on which it is installed.

The rotor consists of a hub and blades, the latter usually made of fibreglass. The blades are designed with an aerodynamical shape to enable rotation when hit by the incident wind. The rotor is connected through a shaft to the many sub-systems assembled inside the nacelle, which is the cabin that hosts the mechanical and electric systems devised to convert the kinetic energy of the wind. The shaft conveys, through a

Zini G., Tartarini P.: Solar Hydrogen Energy Systems. Science and Technology for the Hydrogen Economy. DOI 10.1007/978-88-470-1998-0_5, © Springer-Verlag Italia 2012

Fig. 5.1. Aerogenerators in a wind farm [6]

rotation adapter, the rotational motion of the rotor to the electric generator that converts the mechanical energy in the rotor rotation into electric energy.

The brake system can be electromechanical or hydraulic. It must stop the rotor in case of emergency or maintain the rotational speed within the safety limits. It also functions as a service brake to block the rotation between production sessions.

The rotation multiplier transforms the low number of rotor rotations into a higher angular speed able to drive the electricity production in the generator.

The control system regulates the blade and the rotor positions so that the electric energy conversion is maximised. It also activates the emergency and service brake systems when necessary to make sure that the operation does not exceed the safety limits to damage the structure integrity. A yaw system (like a tail fin or a more sophisticated motorised system) makes sure that the rotor is lined up with the incident wind.

5.2 Mathematical Description of Wind

The wind velocity v at A height z from the ground level can be mathematically modelled with the following equation:

$$v_2 = v_1 \left(\frac{z_2}{z_1} \right)^{\alpha} \tag{5.1}$$

Table 5.1. Roughness coefficients of different land surfaces

Types of land	*Roughness coefficient*
Lake, ocean, hard and smooth land surface	0,01
Land with tall vegetation	0,15
Tall crops, hedges, shrubs	0,20
Dense woodland	0,25
Small city with trees and shrubs	0,30
City with tall buildings	0,40

where v_1 and z_1 are the references of velocity and altitude respectively and α is the roughness coefficient of the surface (as listed in Table 5.1).

The energy distribution of an air flow is evaluated by discretisation of the wind velocities into k classes of width Δv with central values v_i which are average measurements taken over a period of time t_i. The frequencies h_i of each class are calculated with the formula:

$$h_i = \frac{t_i}{\sum_{i=1}^{k} t_i}. \tag{5.2}$$

A frequency distribution is produced in the bar chart in Figure 5.2, in which the bars are interpolated with a *probability density function* (PDF).

The changes of wind velocity in the frequency distribution curve are described by Weibull's probability density functions, given by:

$$h(v) = \left(\frac{k}{c}\right)\left(\frac{v}{c}\right)^{k-1} \exp\left[-\left(\frac{v}{c}\right)\right]^{k} \tag{5.3}$$

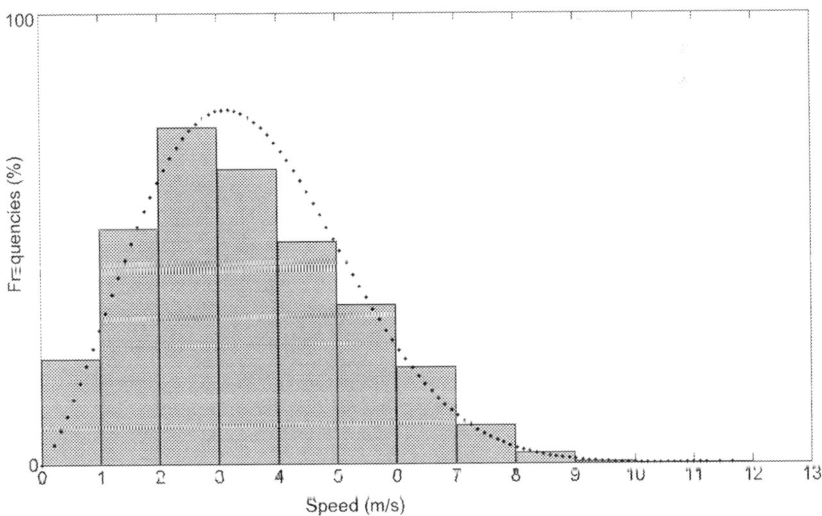

Fig. 5.2. Wind speed frequencies distribution

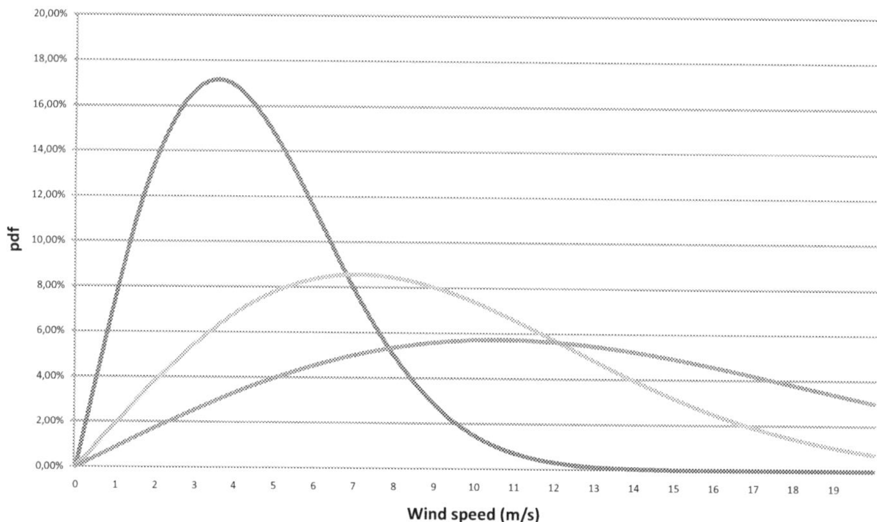

Fig. 5.3. Rayleigh functions according to the three different values of the scale factor c (5, 10, 15)

where c is the *scale factor* which increases with the number of days with high wind speeds. The value of k determines the form of the curve and is therefore known as the *form factor*. It has been observed that, in most locations, the Weibull's function best approximating the frequency distribution is the Rayleigh function, a particular type of Weibull's function with $k = 2$. This function represents the continuous probability density of the module of a two-dimensional vector (i.e. velocity and direction) when these two dimensions are uncorrelated, distributed normally and with the same variance. In a Rayleigh distribution, the days in which the wind speed is below the mean are more than the days when the speed is above the average. The Rayleigh distribution is given by:

$$h(v) = 2 \left(\frac{1}{c} \right)^2 v \exp\left[-\left(\frac{v}{c} \right)^2 \right]. \tag{5.4}$$

Figure 5.3 shows the three different shapes of the Rayleigh function according to the changing values of c.

In most wind farms k shifts in the range of 1.5 to 2.5 while c varies from 5 to 15 m/s.

5.3 Wind Classification

Wind energy systems are designed to function depending on different *wind classes* (Table 5.2). Given the same power rate, the turbines designed to work with winds of

Table 5.2. Wind classes

Class	I	II	III	IV
Average maximum velocity recorded in 50 years for a duration of 10 minutes (m/s)	50	42,5	37,5	30
Average velocity (m/s)	10	8,5	7,5	6

higher classes generally have a larger rotor diameter and a higher tower compared to those operating with lower-class winds.

The characteristics of the wind in a certain location change according to many factors, such as the type and the morphology of the terrain, the height and the period of the year. These elements need to be considered when deciding on the location of a wind farm, along with the possibility that the wind features in the site of interest might change over a medium-long period of time.

Ideally, the wind speed in a location should be determined after a very long period of observation, usually no less than ten year. This time frame is obviously unfeasible, so if the historical data on the local wind speed are not available, the solution is to adopt the *measure, correlate and predict approach* (MCP). First the wind velocity is observed over a period of one year and then the gathered measurements are compared to the historical data of another nearby location. The result of this comparative analysis is then used to estimate the wind velocity of the site of interest for a period longer than one year.

5.4 Mathematical Model of the Aerogenerator

From fluid dynamics, the power of an air flow sweeping across a surface A with a velocity v_1 is given by:

$$P_w = \frac{\rho}{2} A v_1^3 \tag{5.5}$$

where ρ is the specific mass of the air, as a function of the temperature and the pressure of the same air mass. With this equation it is possibile to compute the power and energy probability density functions. The power PDF has a mode speed[1] different from the mode speed of the wind speed PDF since the power varies with the cube of the wind speed. The aerogenerator is designed to collect the most energy possible, therefore the wind turbine design will need to consider the wind power PDF and not the wind speed PDF.

The air density changes according to the pressure and the temperature as per the law of gases. The change of air density according to the elevation represents an important parameter in a correct site design. From measurements taken at sea level, the

---

[1] The *mode speed* is the speed corresponding to the *mode* of the PDF, where the mode is the maximum value assumed by the PDF.

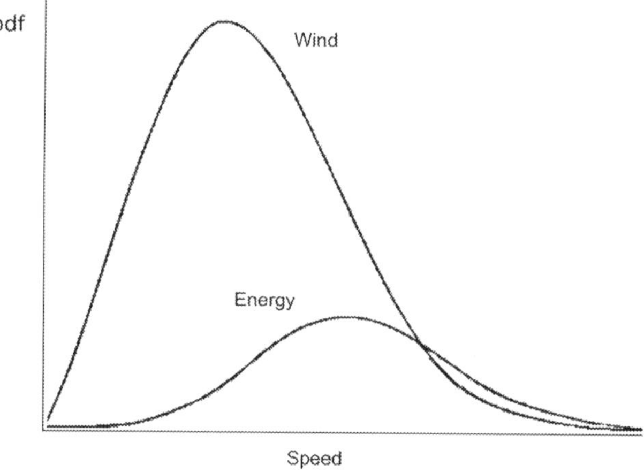

Fig. 5.4. Relation between wind speed and energy distributions

following empirical relation can be considered valid at an altitude under 6000 m:

$$\rho = \rho_0 \, \exp^{\frac{-0.297}{3048} H_m} \tag{5.6}$$

where H_m stands for the altitude of the wind farm site.

The relation in Equation 5.6 can be simplified as:

$$\rho = \rho_0 - 1.194 \times 10^{-4} H_m. \tag{5.7}$$

In an analogous empirical relation, the change in temperature according to height is given by:

$$T = 15.5 - \frac{19.83}{3048} H_m. \tag{5.8}$$

The exploitable mechanical energy P can be expressed with the *power coefficient* c_P:

$$P = c_P P_w = c_P \frac{\rho}{2} A v_1^3. \tag{5.9}$$

The air flow downstream the aerogenerator is slower than that of the incoming wind upstream the turbine since a portion of the kinetic energy of the incident air flow is transferred to the rotational motion of the blades. Assuming that the upstream air flow is uniform at constant speed v_1, the velocity v_2 behind the blades can be approximately considered to be the average between v_1 and a speed v_3 measured at a distance sufficiently far away from the turbine. It can be shown that the maximum theoretical energy is generated when $v_3 / v_1 = 1/3$. In such case, c_P becomes 0.59 and is known as the *Betz's limit*. In reality, however, this value is reduced to between 0.4 and 0.5 due to aerodynamical losses.

The friction and the imperfections in the mechanical couplings can further reduce the exploitable mechanical energy. Other energy losses are caused when the mechanical energy is transferred to the rotor of the electric generator through a shaft and when such energy is converted to electric energy.

The *Bernoulli's equation* is expressed as:

$$\dot{m}\left[\frac{v_2^2 - v_\infty^2}{2} + g\left(z_2 - z_\infty\right) + \int_\infty^2 u\,dp + L' + R'\right] \approx \dot{m}\left(\frac{v_2^2 - v^2}{2} + L' + R'\right) = 0 \quad (5.10)$$

in which $\dot{m}$ is the mass flow rate of the air, v is the wind velocity, L is the work, R is the friction, u is the volume, and p is the pressure.

Assuming that $z_2 = z_\infty$ and $p_2 = p_\infty$, the following is obtained:

$$P = \dot{m}L' = \dot{m}\left(\frac{v^2 - v_2^2}{2} - R'\right). \quad (5.11)$$

According to the equation of continuity applied to the air flow, assuming that $R' = 0$ and introducing the *interference factor a* (between 0 and 0.5) so that $v_2 = v_1\left(1 - a\right)$ and $v_3 = v_1\left(1 - 2a\right)$, the result is:

$$\frac{v^2 - v_2^2}{2} = \frac{v^2}{2}\left[1 - (1 - 2a)^2\right] = 2a\left(1 - a\right)v^2 \quad (5.12)$$

$$\dot{m} = \rho v_1 A_1 = \frac{\pi}{4}\left(1 - a\right)\rho v D^2 \quad (5.13)$$

where D is the diameter of the rotor.

Substituting equations 5.12 and 5.13 with 5.11:

$$P = \dot{m}\frac{v^2 - v_2^2}{2} = \frac{\pi}{2}a(1 - a)^2\rho v^3 D^2. \quad (5.14)$$

The turbine efficiency is evaluated as the ratio between the P in Equation 5.14 and the energy P_{max} contained in an air flow passing through the surface A:

$$P_{max} = \frac{\rho}{2}A v^3 = \frac{\pi}{8}\rho v^3 D^2. \quad (5.15)$$

Therefore the efficiency is given by:

$$\eta = c_P = \frac{P}{P_{max}} = 4a\left(1 - a\right)^2. \quad (5.16)$$

By derivation of η with respect to a and equating the results to zero, the value of a is obtained equal to $1/3$. For such value of a, the corresponding maximum theoretical efficiency is 0.593. The efficiency of the turbine is the previously mentioned energy coefficient c_P.

The *tip-speed ratio* λ (TSR) is a very useful parameter in modelling the behaviour of a wind turbine. It is the ratio between the speed of the blade tip v_t and that of the incident wind v_1, as.

$$\lambda = \frac{v_t}{v_1} = \frac{D\omega}{2}\frac{1}{v_1} \quad (5.17)$$

where D is the diameter of the rotor swept area and ω is the angular velocity of the blades. The term $D\omega/2$ is the tip speed of the blades.

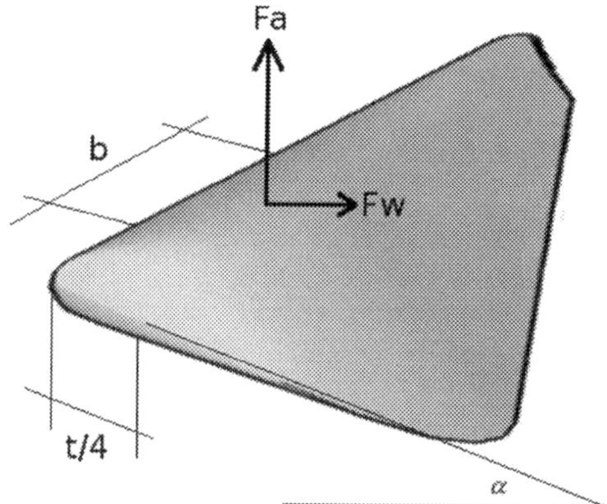

Fig. 5.5. Schematic of the forces acting on the blade

In a rotating mechanical system, the power is given by the product of the torque and the angular velocity. Dividing Equation 5.9 by the angular speed ω, from Equation 5.17:

$$T = \frac{c_P \frac{\rho}{2} A v_1^3}{\omega} = \frac{c_P}{\lambda} \frac{\rho}{2} \frac{D}{2} A v_1^2 = c_T \frac{\rho}{2} \frac{D}{2} A v_1^2 \qquad (5.18)$$

in which:

$$c_T = \frac{c_P}{\lambda} \qquad (5.19)$$

is the *torque coefficient*.

The power P is proportional to the cube of the wind speed while the torque T varies with the square of the speed.

In aerodynamics, the forces operating on the blade profile are the *lifting force F_a* and the *drag force F_w* (see Figure 5.5).

Depending on the angle of *pitch α*, which is the angle formed by the wind direction and by the median section line of the blade profile, F_A and F_W can be expressed as:

$$F_a = c_A(\alpha) \frac{\rho}{2} v_1^2 t\, b \qquad (5.20)$$

$$F_w = c_W(\alpha) \frac{\rho}{2} v_1^2 t\, b \qquad (5.21)$$

in which t is the width and b the length of the blade. F_a acts in the normal direction of the air stream while F_w in the same direction of the drag coefficient c_W.

The values of the coefficients c_A and c_W depend on the blade design and on the pitch angle. The ratio between c_A and c_W is called *glide* (or *lift/drag*) *ratio*.

Since according to Betz the speed v_2 is two-thirds of v_1 and the angular speed on a given point of the radius r is $v(r) = \omega r$, the two speeds combine together, as vectors, to yield the wind velocity $c(r)$ as in Figure 5.6.

The increments dF_A and dF_W operate on the area $t\, dr$ (where t is the breadth of the blade at a given radius r as in Figure 5.7) according to the tangential component

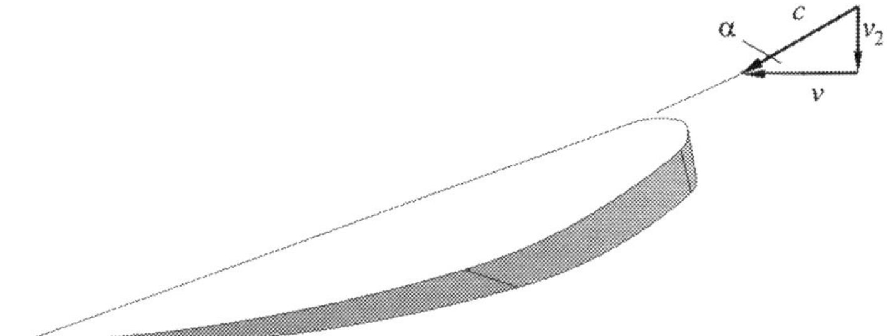

Fig. 5.6. Combination of forces on the blade profile

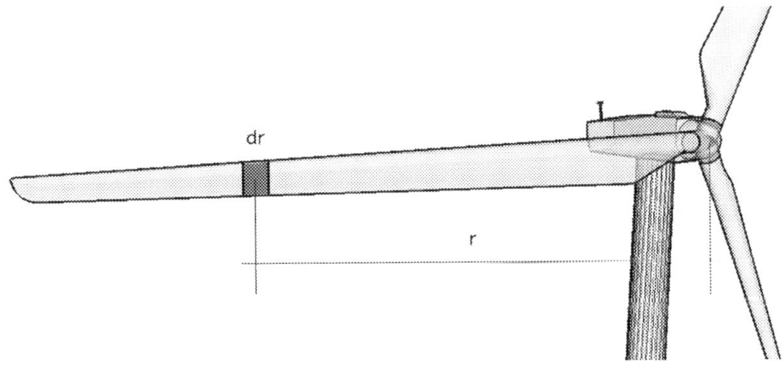

Fig. 5.7. Area for the calculus of the tangential and axial components

dF_t and the axial component dF_a:

$$\begin{bmatrix} dF_t \\ dF_a \end{bmatrix} = \left(\frac{\rho}{2} c^2 t\, dr \right) \begin{bmatrix} c_A \sin(\alpha) - c_W \cos(\alpha) \\ c_A \cos(\alpha) + c_W \sin(\alpha) \end{bmatrix}. \tag{5.22}$$

The integration of the tangential components yields the torque, while the integration of the axial components provide the drag force acting axially on the rotor.

At the tip of a blade with a radius R, the angular speed is given by $v(R) = \omega R$ while the relative wind speed is:

$$c(R) = v_2 \sqrt{1 + \lambda^2}. \tag{5.23}$$

Table 5.3 lists the values of the power coefficient and the torque for different types of rotors.

The aerogenerators with one or two blades (the so-called *fast running* wind turbines) have high power coefficients at high λ but low torque coefficients. Fast-running turbines also tend to be noisier. The best overall performances are currently provided by 3-blade generators, which are nowadays the preferred choice for wind farms.

Decreases of λ cause a reduction of the power coefficient, whereas increases of λ can only raise the power coefficient up to a maximum value that depends on the

Table 5.3. Maximum power and torque coefficients for different types of rotor (at corresponding λ)

Rotor	Power c_P	Torque c_T
4 blade	0.28 (2.3)	0.19 (1.2)
3 blade	0.49 (7)	0.18 (2.3)
2 blade	0.47 (10)	0.07 (7)
1 blade	0.42 (15)	0.04 (14)

design of the blade profile and its optimal pitch angle α. If the tip speed increases and the pitch angle remains the same, the lift force would be reduced, the drag force augmented and the blades would enter into a stall. Figures 5.8 and 5.9 show c_P and c_T for a 3-blade aerogenerator, with a fixed α and when the blades are designed for an optimal $\lambda = 6.5$.

The drag coefficient c_W for a 3-blade turbine as a function of λ is shown in Figure 5.10.

The power P, the torque T and the axial drag force S are calculated by the following equations:

$$P = c_P(\lambda)\, v_1\, \frac{\rho}{2}\, \frac{D^2 \pi}{4}\, v_1^2 \tag{5.24}$$

$$T = c_T(\lambda)\, \frac{D}{2}\, \frac{\rho}{2}\, \frac{D^2 \pi}{4}\, v_1^2 \tag{5.25}$$

$$S = c_W(\lambda)\, \frac{\rho}{2}\, \frac{D^2 \pi}{4}\, v_1^2. \tag{5.26}$$

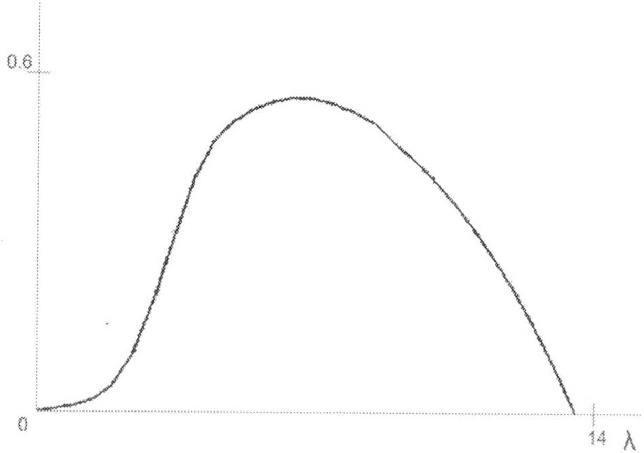

Fig. 5.8. Power coefficient for a 3-blade rotor

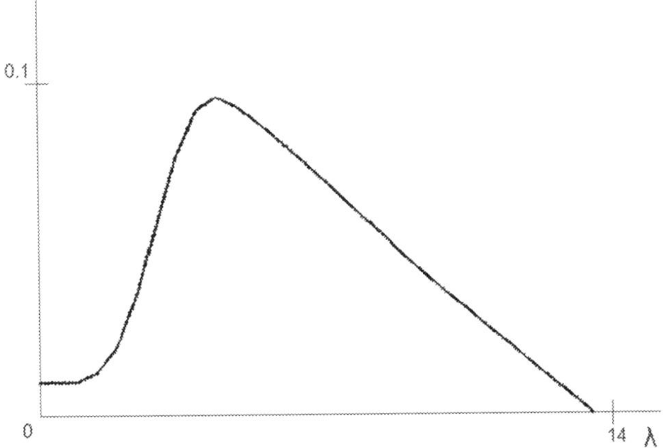

Fig. 5.9. Torque coefficient for a 3-blade rotor

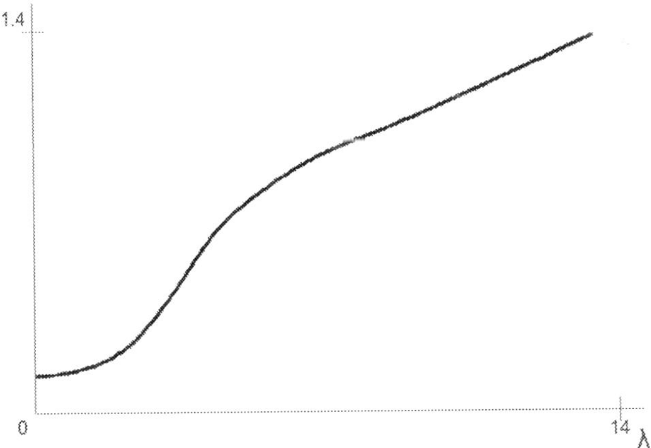

Fig. 5.10. Drag coefficient for a 3-blade rotor

5.5 Power Control and Design

In order to control the power produced by the turbine and prevent the structural over-
load when the wind speed exceeds the design wind speed limits, the pitch angle α
must be adjusted. The blades are controlled by a hydraulic or electric *pitching* system
that increases the blade pitch angle when the power and torque coefficients need to be
reduced. When the pitch angle surpasses a certain limit, the laminar air flow becomes
turbulent, causing the blades to stall and slowing down the rotation.

The curve in Figure 5.11 shows that power cannot be provided when the wind
velocity is lower than the minimum limit, which usually falls between 3 and 4 m/s,
defined as the *cut-in speed*. Between the cut-in speed and the rated speed, the curve
progresses according to the cube of wind speed. When the rated speed is reached,

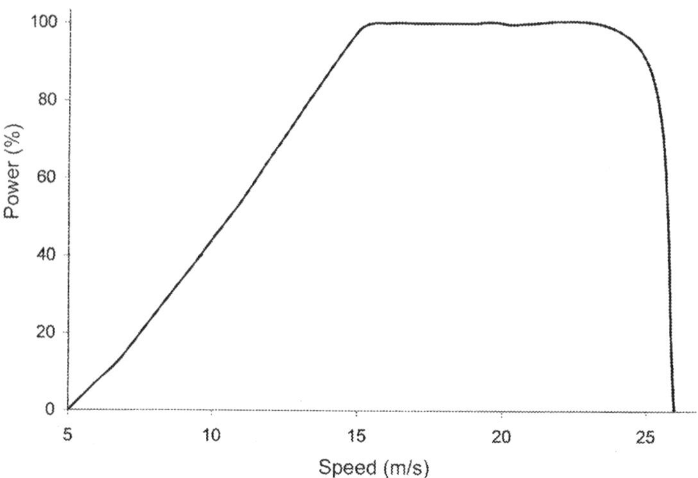

Fig. 5.11. Control of power according to wind speed

the turbine control system keeps the output power at the same level even if the wind speed continues to increase. To protect the structure integrity, the turbine shuts down automatically when the speed surpasses a threshold (*cut-out speed*) which depends on the system design (usually set at around 25 m/s).

The dimensioning of a wind turbine however does not consider the maximum local wind speeds that occur only for a few days per year, but rather the more frequently-observed velocities. Normal wind speeds usually fall in the range of 12 to 16 m/s.

The rotation can be stopped by placing the rotor axis in a position normal to the air flow direction or by using the hydraulic or mechanical brake of the rotor shaft, which is a solution independent from the relative position of the rotor axis with the wind direction. Safety measures and technical regulations request the installation of a second brake system which can also be either hydraulic or mechanic.

As shown in Figure 5.12, the maximum power point (MPP) changes with the wind speed. This MPP can be reached by controlling the rotation speed in order to achieve the best TSR (λ) possible and obtain the highest energy productivity with all meteorological conditions.

In a way similar to how the MPPT logic operates in a photovoltaic system, the turbine control system regulates the rotor speed according to the wind velocity with its peak power tracking technology. Since the power curve has only one apex, finding the solution to the equation which sets $dP/d\omega$ to zero means finding the angular speed that converts the most energy at every wind speed. An automatic electronic control system changes the rotor speed continuously within short intervals: when the amount of energy produced increases or decreases, the system will either decrease or increase the rotor speed in order to maximise energy conversion from the air flow. The rotor speed adjustment also serves the purpose of avoiding mechanical damages

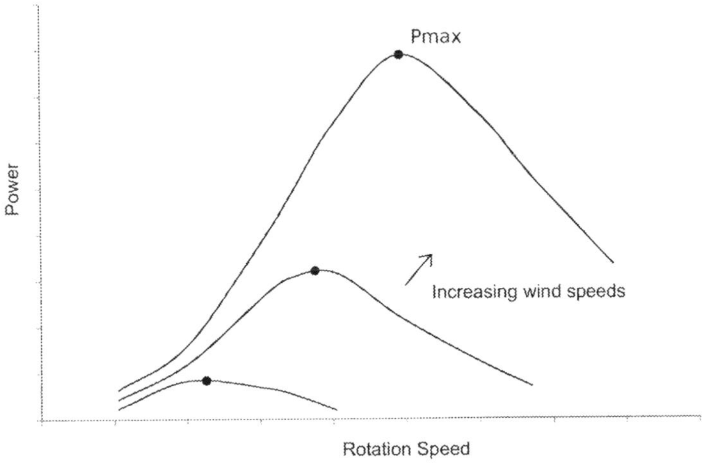

Fig. 5.12. Variation of the maximum power point with the wind speed

to the structure. In this regard, the operating ranges of the turbine are divided into the following:

- *cut-in zone* to avoid turbine functioning at negligible power output;
- c_P *constant zone* the rotor speed is regulated so that c_P remains constant in order to always convert the maximum energy possible;
- *constant power (continuous speed) zone* the turbine maintains a constant rotor speed and a fixed power output to prevent mechanic and electric overcharges;
- *cut-out zone* the turbine operation is stopped when the wind speed exceeds the maximum design limit.

Many other parameters and constraints must also be considered in wind farm design, such as the average wind speeds measured at different heights and the structural reinforcements that need to be able to protect the turbine against the rare and exceptionally strong storms (the *once-in-a-century storms*).

The angular velocity variations caused by the differences between the mechanical and electric powers are given by:

$$J \frac{d\omega}{dt} = \frac{P_m - P_e}{\omega} \tag{5.27}$$

where J is the moment of inertia of the rotor, ω is the rotor angular velocity, P_m is the mechanical power and P_e is the electric power produced by the generator. By the integration of Equation 5.27:

$$\frac{1}{2} J \left(\omega_2^2 - \omega_1^2 \right) = \int_{t_1}^{t_2} (P_m - P_e) \, dt . \tag{5.28}$$

When a rotor has a moment of inertia close to 8000 kg m^2 and needs to reduce its speed from 100 to 95 rpm in 5 s, the power applied by the braking system to the motor will be as high as nearly 800 kW. The brake torque will be as high as to pose considerable mechanical stresses on the structures. To reduce the speed in just 1 s, the required power will be nearly five times higher with a resisting torque very likely to damage the system or, at least, reduce its service life significantly. Since regulating the rotor speed signifies applying accelerations and decelerations that may potentially cause mechanical and electrical damages, the control system must be designed very accurately considering the many different meteorological conditions that might occur in the chosen location.

5.6 Wind Turbine Rating

The power converted from commercial wind turbines ranges from a few kW for stand-alone systems to over 2 MW for large grid-connected plants.

At the moment there is no internationally-agreed standard to rate the turbines as there are many variables determining the power delivered. Since a turbine converts different amount of energy according to different wind velocities, in theory a standard speed should be defined for reference. This however can be difficult to do, since when the incident wind speed is the same, the amount of energy converted can vary depending on the different rotor specifications.

The most widely-adopted standard at the moment by aerogenerator manufacturers is the *Specific Rated Capacity* (SRC), which is calculated as the ratio between the power of the electric generator and the surface area of the rotor. Consequently, a 350/30 turbine has a power of 350 kW with a rotor diameter of 30 m, with an SRC of 0.495 kW/m^2. This index is used in turbine design as it is helpful in correctly dimensioning the mechanical and electric systems as well as in analysing the economics related to the annual energy productivity.

5.7 Electric Energy Conversion

The most commonly used device to convert wind energy to electric energy is the electromagnetic induction asynchronous generator. This device is popular in industrial applications because of its very high reliability and low maintenance cost. The induction generator is hosted inside the nacelle and its axis is fitted to the wind turbine shaft. The wind turbine rotation prompts the induction generator rotor movement which consequently varies the magnetic field concatenating between the rotor and the stator. By electromagnetic induction, an electric field is generated in the stator and electric energy is injected into the grid or stored. Thanks to its functioning characteristic, asynchronous generator is capable of smoothing the frequent and rapid variations of the shaft rotation speed, which is a phenomenon commonly-found in wind turbines. This is why the asynchronous generator is particularly well-suited for wind energy applications.

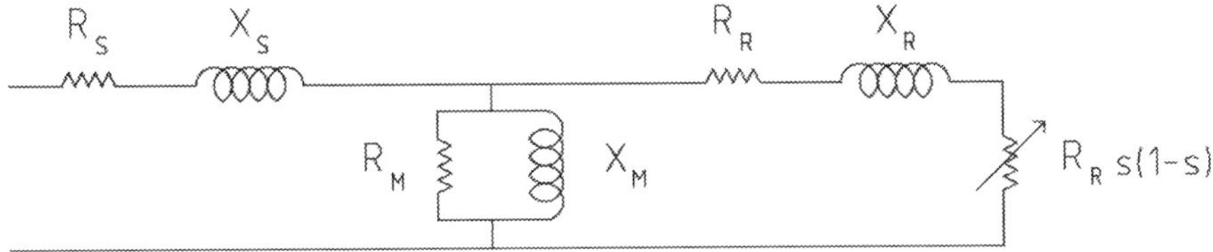

Fig. 5.13. Equivalent circuit of the stator and the rotor with respect to the stator circuit

The power delivered to the grid depends on the *slip* of the asynchronous generator, defined as:

$$s = \frac{N_S - N_R}{N_S} \tag{5.29}$$

where N_S is the synchronous speed at which the speeds of the rotor and of the electromagnetic field are exactly the same, while N_R is the actual rotor speed. The power is generated when the rotor moves slightly faster than the synchronous speed.

The equivalent circuit of the asynchronous generator is shown in Figure 5.13.

R_S and R_R are the resistances of the stator and the rotor respectively while X_S and X_R are the reactances that account for the electromagnetic flow that does not link the rotor to the stator (*leakage flux*). R_M and X_M represent the magnetisation resistance and reactance of dissipative phenomena like parasitic currents and magnetic hysteresis.

The power for every phase is calculated by:

$$P = I_R^2 R_R \left(1 - s\right)/s \tag{5.30}$$

where I_R is the rotor current. Since the system comprises of three phases, the total power is three times the power expressed by Equation 5.30.

R_S and R_R are calculated usually as a percentage of the nominal phase impedances and represent the electric losses during the functioning.

The conversion efficiency of the asynchronous generator can be calculated as:

$$\eta = 1 - 2\left(R_S + R_R\right). \tag{5.31}$$

As an example, if R_S and R_R equal 2%, the performance efficiency becomes 92%, meaning that 8% of the input energy is lost during the conversion process.

For an asynchronous generator, the energy lost during the conversion is released in the form of heat which needs to be removed to prevent overheating and related damages. Cooling can be achieved with the use of air when the amount of energy produced is low, otherwise with water. Compared to air, water is a more efficient cooling agent and it also helps reduce the dimension of the generator and the noise caused by vibration during the operation.

5.8 Calculation Example

The following calculations are based on the example of a location with an air density $\rho = 1.1$ kg/m^3, an average annual wind speed $c = 11$ m/s and a wind turbine with a performance ratio $\eta = 0.5$ and a diameter $D = 30$ m.

Theoretically, the maximum obtainable power is calculated as:

$$P_{max} = \frac{\rho}{2} A c^3 = \frac{\pi}{8} \rho c^3 D^2 = 517 \text{ kW.} \tag{5.32}$$

In reality the power output is about half of the above capacity without considering the electromechanical conversion:

$$P = \eta \times P_{max} = 0,5 \times 517 = 258 \text{ kW.} \tag{5.33}$$

In this case, the power P' per unit surface area is given by:

$$P' = P/A = 258/706 = 365 \text{ W/m}^2 \tag{5.34}$$

As a result, the maximum estimated annual energy output is around 2.26 MWh.

5.9 Environmental Impact

The installation of wind turbines often produces noticeable environmental impacts on the surrounding areas. The first concern rises from the visual impact on the unpopulated natural landscape whose scenery integrity is often considered to be valuable and needs to be properly guarded. The presence of the turbines can also disturb the living habits and the usual migration routes of the aviary population and even cause deaths when the birds accidentally hit the blades. Another aspect that is often considered a downside is the noise generated by the turbine operation. However, turbines in reality are not as loud as what is generally perceived. For example, the noise of a turbine with a capacity of 600 kW is measured at 55 dB at a distance of 50 m and 40 dB at 205 m, no louder than the noise coming from a normal factory in comparison. The volume of the noise is usually constant and only peaks when the yaw is being adjusted following the changing wind direction.

Another issue is related the electromagnetic interferences that turbine operations pose on the nearby electric devices, which is an effect caused by the electric and constructive characteristics of the blades. It is possible that the local reception of electromagnetic signals be disturbed by the presence of one or more turbines.

These factors all need to be taken into account in the planning of the position of the wind farms. This is why public authorities request detailed screenings and evaluations on the environmental impacts before releasing construction permits, extending therefore the time-frame of actual wind farm commissioning.

References

1. Ayotte K W (2008) Computational modelling for wind energy assessment. Journal of Wind Engineering and Industrial Aerodynamics 96:1571–1590
2. Betz A (1926) Windenergie und ihre Ausnutzung durch Windmühlen. Vandenhoek und Rupprecht, Göttingen
3. Burton T, Sharpe D, Jenkins N, Bossanji E (2001) Wind Energy Handbook. John Wiley & Sons Ltd, Chichester
4. Delarue Ph, Bouscayrol A, Tounzi A, Guillaud X et al (2003) Modelling, control and simulation of an overall wind energy conversion system. Renewable Energy 28:1169–1185
5. Gasch R, Twele J (2007) Windkraftanlagen. Teubner, Stuttgart
6. Billy Hathorn (photographer) [CC-BY-SA-3.0 (www.creativecommons.org/licenses/by-sa/3.0)], via Wikimedia Commons, http://commons.wikimedia.org/wiki/File:Wildorado_Wind_Ranch,_Oldham_County,_TX_IMG_4919.JPG
7. Patel M R (1999) Wind and Solar Power Systems. CRC Press, Boca Raton
8. Schmitz G (1955-1956) Theorie und Entwurf von Windrädern optimaler Leistung. Z. d. Universität Rostock
9. Stiebler M (2008) Wind Energy Systems for Electric Power Generation. Springer-Verlag, Berlin Heidelberg

Other Renewable Energy Sources for Hydrogen Production

Other sources of renewable energy are available for hydrogen production: hydro-electricity, tidal energy, wave energy, ocean thermal energy, solar thermal energy and biomasses can all be considered as alternatives to wind energy and solar photovoltaic energy. The advanced production procedures based on the reaction of sunlight with organic and inorganic materials can be another viable technique to obtain hydrogen.

6.1 Solar Thermal Energy

The energy irradiated from the sun can be converted in thermal energy. A *low temperature solar thermal* plant employing temperatures not higher than 80°C can be devised for building and water heating systems. Such low temperatures, however, are not suitable for hydrogen production as higher temperatures are needed.

In *concentrated solar thermal* plants, higher temperatures (over 120°C) can be reached by concentrating sunlight with mirrors to a boiler and taking advantage of thermodynamic cycles like the *Rankine cycle*. Efficiencies are in the range of 15–20% with temperatures reaching up to 200°C.

The Rankine cycle has been used in many applications in the past (i.e. in rail-road steam engines) as well as today (i.e. in thermoelectric energy plants). An ideal Rankine cycle can be represented by the *Temperature-Entropy T-S diagram* (Figure 6.1). In the transformation between states 1 and 2, the fluid undergoes an isoentropic pressurisation, followed by an isobaric heating (within 2–2') and a dry-saturated vaporisation (2'-3). In the transformation from state 3 to state 4, the dry saturated vapour expands through a turbine with isoentropic expansion and provides the work that will be converted in electric energy. Finally, between 4 and 1, the wet vapour fluid with a low pressure is taken to the saturated liquid phase by a condenser. The thermovector fluids can be water or low-boiling organic fluids like halogenated hydrocarbons.

The Rankine cycle efficiency is defined by the following relation (Figure 6.2):

$$\eta = \frac{\dot{W}_{turbine} - \dot{W}_{pump}}{\dot{Q}_{in}} \qquad (6.1)$$

Zini G., Tartarini P.: Solar Hydrogen Energy Systems. Science and Technology for the Hydrogen Economy. DOI 10.1007/978-88-470-1998-0_6, © Springer-Verlag Italia 2012

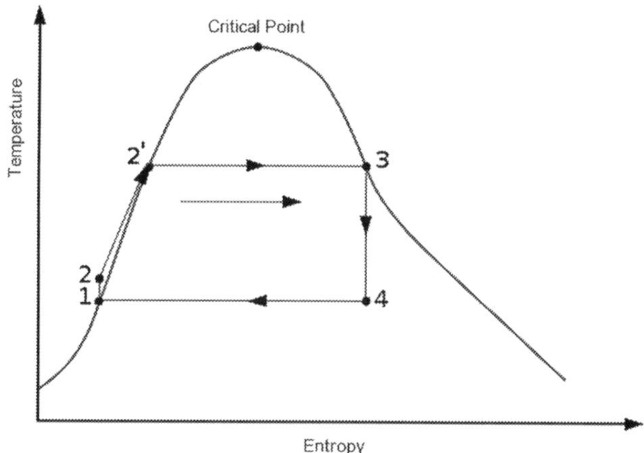

Fig. 6.1. Rankine cycle in the *T-S* diagram

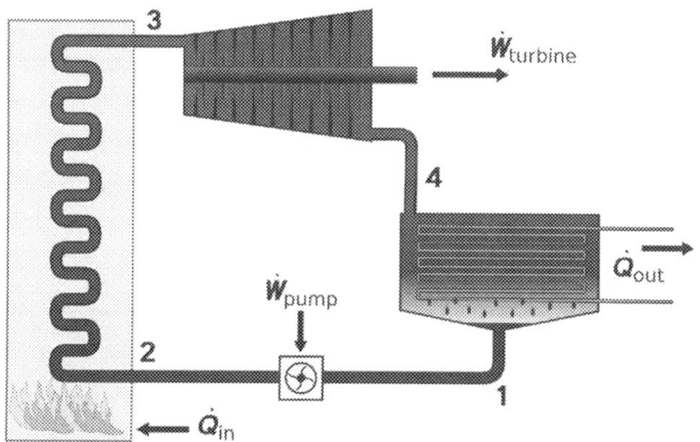

Fig. 6.2. Rankine cycle work and heat power flows [1]

where $\dot{W}$ is the power provided by the turbine or used by the pump and $\dot{Q}_{in}$ is the thermal power to the system. The turbine generates electricity to be used in the electrolytic production of hydrogen.

Concentrated solar thermal plants can be engineered to reach very high temperatures so that hydrogen can be produced through *direct thermolysis* of water. At ambient temperatures and pressures, only one molecule in 10^{14} dissociates by the effect of heat. At a temperature as high as 2200 °C, about 8% of the water molecules are decomposed, and the percentage is raised to 50% when the temperature climbs to 3000 °C. However, the currently available technology suffers from a very high thermal stress associated to direct water thermolysis, and the reliability of these plants still needs to be improved by further developments in material science and in thermal plant technology.

6.2 Hydroelectric Energy

Hydroelectric energy is probably the most mature among the existing renewable energy technologies, with a history of utilisation that dates centuries back. It functions on the principle that the potential energy possessed by still water in a reservoir or the kinetic energy of flowing water can be easily and efficiently converted into electricity.

In the case of a reservoir, the power is expressed by:

$$P = \eta \rho g \Delta z \frac{dV}{dt} \tag{6.2}$$

where η is the efficiency that considers the friction losses, ρ is the water density, g is the gravity constant, Δz is the difference in height between the pipe inlet and the rotating turbine and dV/dt is the water volumetric flow rate.

The design of a hydroelectric plant needs to be based on the volumetric flow rate of the hydrographic catchment area during the course of a full year and possibly longer.

This technology has many advantages as well as drawbacks, which are mainly connected to its severe environmental impact. Along the years the construction of a dam has caused many damages to the surrounding wildlife and cultures. Local residents have been forced to relocate and even entire communities have been wiped out by the sudden flooding caused by dam breakages or natural disasters. Another impact comes from the emission of climate-changing gases. Many reservoirs have been created artificially by blocking natural water flows to cause permanent flooding over a large extended surface. When the woods in the area are immersed in the water, they release a high quantity of CO_2 or CH_4 after their death. Such gases can be released continuously when the water moves up and down to flood other portions of terrain previously covered by vegetation. Some scholars maintain that this periodical release of CO_2 can be compared to the same emission coming from the traditional fossil fuel.

Additionally, the productivity of hydroelectric plants can be threatened by climate changes, if prolonged periods of draught reduce the water volumetric flow rates.

6.3 Tidal, Wave and Ocean Thermal Energy Conversions

Producing hydrogen by converting the potential energy in the tides and the waves with electrolysis can be a promising option, since the two main elements involved, the energy and the water, are readily available on site.

There are several designs that can be considered. At the moment, hydrogen production is feasible on large maritime off-shore platforms which are also engaged in maricultural activities or operate as freight logistic hubs. Hydrogen then can be transported to mainland or even be used to supply hydrogen-fuelled ships in the future.

Tides are generated as the result of the interaction among the Earth, the Moon, the Earth's rotation and depend on the local geography. The *tidal energy* that can be retrieved is more predictable than other renewable energy sources. Recent advances in turbine technology are reducing the costs and increasing the market share of this energy source. The energy calculations are similar to those developed for wind energy due to the analogies between turbine construction and fluid dynamics.

Wave energy is created when the wind energy is transferred to the sea surface. Such energy can be exploited by absorbers or buoys that interact with the oscillating motions of the waves. Practical applications are still under development.

Ocean thermal energy conversion (OTEC) uses a low-pressured Rankine cycle turbine to take advantage of the temperature differences between the waters on the ocean surface and those at deeper levels. The use of this technology to produce hydrogen though is still likely to be minimal at least in the next few decades.

6.4 Biomasses

Biomasses, intended as a renewable energy source, refer to all organic products from plants or animals that have not been fossilised and are available for energy uses as scraps and wastes or as the product of intentional farming.

Plants are *autotroph organisms* since they are capable of providing their own nutrients by storing solar energy in the form of complex organic compounds (mainly carbohydrates) from simple inorganic molecules by *photosynthesis*. All other living creatures are *eterotroph organisms* since they must feed on plants to receive nutrients and, therefore, energy.

One of the processes that can be employed is *direct gasification* in a fluidized bed reactor, where biomasses are converted by heat and by a partial reaction with O_2 into H_2, CO, CH_4, CO_2, steam and other hydrocarbons. Hydrogen must be cooled down and purified from sulphured residuals. Another method is the *fast pyrolysis* where biomasses are degraded by heat to generate H_2, CO, CO_2, hydrocarbons, steam, acids, bases and solid residuals. By carefully selecting the right bacteria and substrates, it is also possible to obtain hydrogen from anaerobic fermentation instead of from the usual methane.

One of the significant advantages of this application is that the wastes are used to provide energy instead of being stored in landfills. The downside is that diverting agricultural activities from food to energy production can reduce food availability and raise its cost. Plus, increasing the land use for energy production can augment the risk of deforestation, which has a direct impact on the carbon cycle and on the climate change. Finally, using high-impact agricultural techniques such as fertilisers, pesticides, machines and irrigation can even wipe out the benefits of employing biomasses to reduce environmental damages.

References

1. Ainsworth A [CC-BY-SA-3.0 (www.creativecommons.org/licenses/by-sa/3.0/)], via Wikimedia Commons, http://commons.wikimedia.org/wiki/File:Rankine_cycle_layout.png
2. Baker A C (1991) Tidal power, Peter Peregrinus Ltd, London
3. Baykara S Z (2004) Experimental solar water thermolysis. International Journal of Hydrogen Energy 29 (14):1459–1469
4. Bruch V L (1994) An Assessment of Research and Development Leadership in Ocean Energy Technologies. SAND93-3946. Sandia National Laboratories: Energy Policy and Planning Department
5. Bolton J R (1996) Solar photoproduction of hydrogen: a review. Solar Energy 1 (57):37–50
6. Frey M (2002) Hydrogenases: hydrogen-activating enzymes. ChemBioChem 3 (2–3): 153–160
7. Herbich J B (2000) Handbook of coastal engineering. McGraw-Hill Professional, New York
8. Mauseth J D (2008) Botany: An Introduction to Plant Biology, 4th ed. Jones & Bartlett Publishers, Sudbury
9. Mitsui T, Ito F, Seya Y, Nakamoto Y (1983) Outline of the 100 kW OTEC Pilot Plant in the Republic of Nauru. IEEE Transactions on Power Apparatus and Systems PAS-102 (9): 3167–3171
10. Nonhebel S (2002) Energy yields in intensive and extensive biomass production systems. Biomass Bioenergy 22:159–167
11. Vignais P M, Billoud B and Meyer J (2001) Classification and phylogeny of hydrogenases. FEMS Microbiol Rev. 25 (4): 455–501

7

Hydrogen Storage

Storage is the key to make renewable energies become more reliable. There are many types of technologies available at the moment but none has yet claimed the dominance. For the next few decades or even a longer term, the method of storage will still very likely be a mixture of different technologies, depending on individual applications. The use of hydrogen can be one of the most interesting options worth considering. This chapter examines several different hydrogen storage technologies, from the most traditional to the more advanced, yet feasible, proposals.

7.1 Issues of Hydrogen Storage

As demonstrated in the previous chapters, it is highly beneficial to the environment and technically feasible to use electrolysis to separate water into hydrogen and oxygen gases and then recombine them to generate electricity. While the procedure to procure hydrogen from renewable sources can be considered technically mature, the storage of hydrogen still presents many challenges to overcome.

The traditional methods like compression and liquefaction entail drawbacks that compromise their effectiveness and large-scale deployment. For example, compression storage delivers low storage density and operates on high working pressures that result in high management costs and serious safety risks. Liquefaction, on the other hand, requires a large quantity of energy seriously reducing the final efficiency of the system. Moreover, the continuous evaporation of the liquefied hydrogen from the tank also limits the use of this technology only to applications where the consumption rate is fast and high operating costs are accepted.

An efficient storage technology is characterised by a high energy density stored either in terms of weight (expressed in *gravimetric capacity*, the ratio between the energy stored and the total mass of the storage system), or in terms of volume (expressed in *volumetric capacity*, the ratio between the energy stored and the total volume of the storage system).

Zini G., Tartarini P.: Solar Hydrogen Energy Systems. Science and Technology for the Hydrogen Economy. DOI 10.1007/978-88-470-1998-0_7, © Springer-Verlag Italia 2012

In order to set the development time frame for research organizations around the world, the United States Department of Energy (DOE) has established 6% (2 kWh/kg) as the goal for the gravimetric capacity to reach by the year 2010 and 9% (3 kWh/kg) by 2015. It is to be noted that these objectives are not binding, especially for stationary applications where weight is generally not an issue. Rather, they represent an important indication of the objectives that current research needs to achieve in order to develop a hydrogen storage system capable of completely replacing the present fossil fuel regime.

Hydrogen storage encounters fewer restrictions in stationary uses as the weight and the size of the material do not present any concern, but these aspects can complicate non-stationary uses. A technology capable of resolving such tradeoff will be the answer to the storage problems.

As previously mentioned, two currently available storage methods are compression and liquefaction, while other new technologies still under development include the uses of carbon structures, nanotechnologies, metallic or chemical hydrides and other innovative ideas. Physisorption of hydrogen in carbon structures such as nanotubes and activated carbons is an interesting method. The features of this type of technology include low operation pressure, low safety risks, high charging-discharging speeds and complete reversibility without causing any hysteresis phenomenon. However, the temperatures need to be very low in order to obtain acceptable gravimetric and volumetric capacities. The use of hydrides can be complicated in terms of heat management, meaning that the heat needs to be removed or supplied during the charging and discharging cycles in a vacuum so that the hydrides do not come in contact with air. Nonetheless, this technology contains high volumetric capacity and holds many worth-noticing features while other new methods are still in an early phase of study. Some groups of hydrides seem to be able to operate nearly at ambient temperature and pressure, while other hydride complexes, such as $LiBH_4$, have demonstrated good volumetric and gravimetric capacities and seem to be considered suitable for non-stationary applications. All these options will be discussed in the next sections.

7.2 Physical Storage

7.2.1 Compression Storage

Currently, compression storage is the simplest technology for hydrogen storage. Considering the low density of hydrogen, storage must be performed under high pressures (from 250–300 up to 700 bar) or in large volumes.

Low pressures can be applicable to large-scale stationary uses, since the reduced pressure can be compensated by the high-volume of large storage tanks. In non-stationary applications, however, it is necessary to accept a compromise between volume and weight reductions. A volume reduction is feasible only at higher pressures and with low weights.

Compared to methane, hydrogen compression storage usually requires a volume 3 times the volume of methane with an amount of specific energy (in MJ/kg) much higher than what is used for methane compression. Hydrogen also needs higher compression pressures due to its lower volumetric energy density.

Commonly used compressors include reciprocating pistons, rotary compressors, centrifugal and axial turbines. Such compressors must be constructed with compatible materials suitable for the contact with hydrogen. The storage tanks are usually made of aluminium reinforced with fibreglass or polymeric materials with carbon fibres, permitting a gravimetric density around 2–5%.

7.2.1.1 Modelling

The typical compressors used in residential settings are either single-stage or two-stage compressors. The ideal work of compression $L_{comp,id}$ of a single-stage compressor conducting a *polytropic compression*[1] can be described as:

$$L_{comp,id} = \frac{m\,R\,T_{in}}{m-1}\left[1 - \left(\frac{p_{out}}{p_{in}}\right)^{\frac{m-1}{m}}\right] \qquad (7.1)$$

where m is the exponent of the polytropic compression, T_{in} is the temperature of the gas entering the compressor and p_{in} and p_{out} are the pressures at the input and the output points of the compressor. The temperature of the gas T_{out} at the outlet can be obtained through a mathematical calculation from the expression of the polytropic process and from the law of perfect gas:

$$p_{in}v_{in}{}^{m} = p_{out}v_{out}{}^{m} \rightarrow$$

$$p_{in}\left(\frac{RT_{in}}{p_{in}}\right)^{m} = p_{out}\left(\frac{RT_{out}}{p_{out}}\right)^{m} \rightarrow$$

$$p_{in}{}^{(1-m)}T_{in}{}^{m} = p_{out}{}^{(1-m)}T_{out}{}^{m} \rightarrow$$

$$T_{out} = T_{in}\left(\frac{p_{in}}{p_{out}}\right)^{\frac{1-m}{m}}. \qquad (7.2)$$

The power effectively absorbed by the compressor P_{comp} is:

$$P_{comp} = \frac{\dot{n}_{gas}L_{comp,id}}{\eta_{comp}} \qquad (7.3)$$

where $\dot{n}_{gas}$ is the molar flow rate of the gas passing through the compressor and η_{comp} is the efficiency of the compressor.

For stationary applications it is more convenient to use large-volume tanks to maintain low storage pressures and energy consumption. For non-stationary uses, instead, it is necessary to adopt smaller tanks but with the highest pressure possible.

---

[1] A thermodynamic transformation is defined as *polytropic* when it follows the law pv^{γ} = constant, in which γ is the characteristic exponent (or the characteristic number) of the polytropic.

Given an ideal gas, the pressure p_s inside the storage tank equals:

$$p_s = \frac{n\,R\,T_s}{V} \tag{7.4}$$

where n is the number of moles of the gas in the storage tank, T_s is the tank temperature and V is the tank volume. The pressure p_s therefore can be calculated from T_s and n, which depends on the molar flow difference between the inlet gas $\dot{n}_{in,s}$ and the outlet gas $\dot{n}_{out,s}$. The temperature T_s is also influenced by the temperature of the gas entering the tank $T_{in,s}$ and by the exiting gas temperature $T_{out,s}$. Therefore, the calculation of p_s requires a system of three non-linear equations for the three unknowns: p_s, T_s and $T_{in,s}$.

The first equation is:

$$T_{in,s} = T_{in,c}\left(\frac{p_{in}}{p_{in,s}}\right)^{\frac{1-m}{m}} \tag{7.5}$$

in which $T_{in,c}$ is the temperature of the gas entering the compressor, which for simplification is assumed to be the same as the gas leaving the electrolyser, ignoring the temperature drop that occurs when the gas passes through the pipe connecting the electrolyser and the compressor.

The second equation is given by the perfect gas law in which the time integral of the mole quantity of the gas n is replaced by n_{in}, the number of moles of the gas in the initial conditions:

$$n = n_{in} + \int_0^t \left(\dot{n}_{in,s} - \dot{n}_{out,s}\right) d\tau. \tag{7.6}$$

The second equation therefore becomes:

$$p_s = \left[n_{in} + \int_0^t \left(\dot{n}_{in,s} - \dot{n}_{out,s}\right) d\tau\right]\frac{R\,T_s}{V}. \tag{7.7}$$

The thermal equilibrium in the tank is expressed by:

$$\dot{Q}_{in} = \dot{Q}_{store} + \dot{Q}_{loss} \tag{7.8}$$

in which the thermal power entering the tank is:

$$\dot{Q}_{in} = \dot{n}_{in,s}\,c_{p,gas}\left(T_{in,s} - T_s\right) = C_{in}\left(T_{in,s} - T_s\right) \tag{7.9}$$

and the thermal power stored in the tank is:

$$\dot{Q}_{store} = C_s\frac{dT_s}{dt} \tag{7.10}$$

while the thermal losses to the environment are given by:

$$\dot{Q}_{loss} = \frac{\left(T_s - T_a\right)}{R_t} \tag{7.11}$$

where $c_{p,gas}$ is the specific heat of the gas stored in the tank, C_{in} is the heat capacity of the entering gas flow, T_a is the ambient temperature (in K), R_t is the thermal resistance of the tank and C_s the heat capacity of the entire tank.

Since the thermal capacity of the tank wall is lower than that of the gas contained within, the following relation can be considered valid:

$$C_s \cong n\,c_{p,gas}.\tag{7.12}$$

By combining equations (7.8) and (7.12), the third equation of the non-linear system is:

$$T_s = T_{ini} + \int_0^t \left[\frac{C_{in}\left(T_{in,s} - T_s\right)}{C_s} + \frac{(T_a - T_s)}{\tau_t} \right] d\tau \tag{7.13}$$

where T_{ini} is the temperature of the tank in the initial conditions, which can be assumed to be the same as the ambient temperature and τ_t is the time constant of the tank, equal to the product $R_t\,C_s$.

The non-linear system which allows the calculation of the tank pressure therefore becomes:

$$\begin{cases} T_{in,s} = T_{in,c}\left(\dfrac{p_{in}}{p_{in,s}}\right)^{\frac{1-m}{m}} \\[2ex] p_s = \left[n_{in} + \int_0^t \left(\dot{n}_{in,s} - \dot{n}_{out,s}\right) d\tau \right] \dfrac{RT_s}{V} \\[2ex] T_s = T_{ini} + \int_0^t \left[\dfrac{C_{in}\left(T_{in,s} - T_s\right)}{C_s} + \dfrac{(T_a - T_s)}{\tau_t} \right] d\tau. \end{cases} \tag{7.14}$$

7.2.1.2 Dimensioning Example

Gas storage tanks are made of steel in a cylindrical form. The thickness of the wall of a hydrogen tank is 5 mm with a storage volume of $0.4\ \text{m}^3$ while an oxygen tank wall has the same thickness but a storage capacity of $0.2\ \text{m}^3$. The length of both tanks, excluding the thickness of the plates at the ends, can be calculated by this formula:

$$l = \frac{V}{\frac{\pi d^2}{4}}.\tag{7.15}$$

Ignoring the thermal bridges which correspond to the passage from cylindrical to plain geometry, these two tanks are two pressurised hollow cylinders with two round plates enclosing both ends. Their thermal resistance can be calculated by the following:

$$R_t = R_{lat} + R_{ext}\tag{7.16}$$

where:

$$R_{lat} = R_{conv.in,lat} + R_{cond,lat} + R_{conv,out,lat} =$$

$$\tag{7.17}$$

$$\frac{1}{h_{in}\pi d l} + \frac{\ln\frac{D}{d}}{2\pi\lambda l} + \frac{1}{h_{out}\pi D l}$$

and:

$$R_{ext} = 2\left(R_{conv.in,ext} + R_{cond.ext} + R_{conv.out,ext}\right) =$$

$$2\left(\frac{1}{h_{in}\,\frac{\pi d^2}{4}} + \frac{\frac{(D-d)}{2}}{\lambda\,\frac{\pi d^2}{4}} + \frac{1}{h_{out}\,\frac{\pi d^2}{4}}\right). \tag{7.18}$$

In equations (7.17) and (7.18):

- R_{lat} is the total thermal resistance of the lateral walls of the cylinder whose internal diameter is d and external diameter is D;
- R_{ext} is the thermal resistance at both ends of the tank enclosed by the two round plates with diameter d;
- $R_{con.in,lat}$ is the internal convective resistance of the lateral surface;
- $R_{cond.lat}$ is the conductive resistance of the lateral surface;
- $R_{conv.out,in}$ is the external convective resistance of the lateral surface;
- h_{in} is the coefficient of the internal convection which can be raised to 200 W/m^2 K in these conditions;
- d is the internal diameter of the cylinder;
- l is the length of the cylinder;
- D is the external diameter of the cylinder;
- λ is the thermal conductivity of the steel layer, which is measured to be around 50 W/m K for most types of steel;
- h_{out} is the external convection coefficient, which can be assumed to be 5 W/m^2 K in these conditions;
- $R_{conv.in,ext}$ is the internal convective resistance of the plates installed on both ends;
- $R_{cond.ext}$ is the conductive resistance of the steel layer of the round plates;
- $R_{conv.out,ext}$ is the external convective resistance of the plates.

7.2.2 Liquefaction Storage

Liquefaction storage can compensate for the low energy density of compressed hydrogen. It allows the stored hydrogen volumetric density to reach 50 kg/m^3 with a gravimetric density close to 20%. However, the very low temperatures of liquefaction can present some problems, as it is difficult to maintain such low temperature against all thermal losses occurring in the storage vessels. Since the storage temperature is close to the boiling point of hydrogen, even a very small heat exchange with the environment can cause evaporation inside the tank with the need to vent gaseous hydrogen to avoid internal over-pressurisation.

Due to the low critical point of hydrogen, the energy consumed by the compressor during refrigeration combined with the thermal losses can make the cost of liquefaction much too high. For example, the H$_2$ liquefaction cost is 3.23 kWh/kg while the cost of N$_2$ liquefaction is only 0.21 kWh/kg. Around 30% of the energy consumed (in terms of the lower heating value of hydrogen) is needed for the liquefaction process, while only 4% is needed for the compression. This is an issue in small-scale non-stationary applications, particularly in automotive uses, as in order to maintain liquefaction, energy is consumed even when the vehicle remains still.

From an engineering point of view, the tanks should preferably take the form of a sphere to guarantee the lowest surface to volume ratio. Furthermore, to minimise the thermal exchanges of conduction, convection and radiation, the tanks are built with an inner and outer vessel with a gap either kept in vacuum or filled with liquid nitrogen at 77 K.

The classical industrial liquefaction process is based on the *Joule-Thompson effect*, that occurs when a gas forced through a valve changes temperature after expansion in adiabatic conditions.

A gas is first compressed at ambient temperature then cooled in a heat exchanger. It is then released through a throttle valve where it undergoes the Joule-Thompson effect that causes the liquefaction of a part of the gas. The gas that remains after the partial liquefaction re-enters in the cycle at the heat exchanger.

The basic *Linde's cycle* works with different types of gases, such as nitrogen, which cools down at ambient temperature when undergoing an isenthalpic expansion. Hydrogen, on the contrary, heats up in an isenthalpic expansion. This is why in order for hydrogen to cool upon expansion, its temperature must be below its inversion temperature[2]. To reach such temperature, a pre-cooling to 78 K is performed before hydrogen is sent to the throttle valve. Under the inversion temperature, the internal interactions of the H_2 molecules cause the gas to do work when it is expanded. Some versions of the Linde's process use liquid nitrogen to pre-cool the hydrogen before it passes through the expansion valve; the nitrogen is then retrieved and reused in the process.

An alternative to the Linde's cycle is the *Claude's cycle*, in which some of the compressed hydrogen is diverted to an engine to undergo an isenthalpic expansion.

The liquefaction temperature of hydrogen is 20.28 K. At liquefied conditions, hydrogen is nearly 100% in the form of para-hydrogen, while at ambient temperature the distribution becomes 25% para and 75% ortho. The conversion from ortho to para-hydrogen releases heat that causes hydrogen boil-off losses. Hydrogen needs to be in its para form for long-term storage, rather than in the ortho form. This must be accomplished by pre-treatment of hydrogen gas with catalysts that perform such conversion before hydrogen is liquefied.

7.2.3 Glass or Plastic Containments

Special glass micro spheres with a diameter between 25 and 500 μm can be used to store hydrogen: when the glass is heated up to 200 and 400 °C and pressurized with several tens of MPa, it becomes permeable to hydrogen (with the highest limit being around 340 MPa, the rupture point of the spheres). When the pressure and the temperature return to normal, hydrogen is sequestered inside the spheres. When hydrogen needs to be released the spheres are heated up again while the system maintains the

---

[2] At a given pressure, a non ideal gas has an *inversion temperature* above which the expansion of the gas in an isenthalpic transformation causes a temperature increase, while below such inversion temperature an expansion causes a temperature decrease.

normal pressure or becomes under-pressurised. The spheres can also be crushed to release their content.

The performance efficiency of this process depends on the hydrogen pressure and the temperature, as well as on the volume, the dimension and tthe chemical composition of the spheres. Since it is an intrinsically safe method, it can be adopted fairly easily in non-stationary applications.

Another solution involves using small plastic spheres filled with NaH that are placed inside a water reservoir with a grinding device. When hydrogen is needed to provide energy, a control system proceeds to grind the plastic spheres to release the content in the water. The chemical reaction that follows is:

$$NaH + H_2O \rightarrow NaOH + H_2 \tag{7.19}$$

with the release of hydrogen and sodium hydroxides. This procedure is integrated into a more complex system capable of retrieving sodium hydroxides for a later re-conversion to NaH. Stored hydrogen density can reach 4.3% by weight or 47 kg/m^3 by mass.

7.3 Physical-Chemical Storage

7.3.1 Physisorption

The atoms on the surface of a material are not completely surrounded or enclosed like the atoms inside the same material, therefore they are freer to interact with other atoms or molecules present in the surrounding environment. These atoms belonging to the surface of the material and other atoms or compounds in the outer environment can come into contact and form bonds. *Adsorption* is the surface phenomenon that occurs between the *adsorbate* atoms, ions or molecules that form bonds with the atoms of the surface of the *adsorbent*.

When the nature of these bonds belongs to Van der Waals' forces (with an adsorption enthalpy of 20 kJ/mol, the same dimension as the enthalpy of condensation), the phenomenon is called *physical adsorption* or *physisorption*. The enthalpy is not sufficient to break the bonds of the original adsorbate, therefore the adsorbed molecules can maintain their nature although their three-dimensional shape can be distorted by the forces of attraction exerted by the atoms belonging to the lattice of the adsorbent (Table 7.1).

Physisorption is a very fast process. It does not need activation energy and the bonds thus formed, being non-specific, are not particularly selective between the adsorbates and the adsorbents. The easiest adsorbed gases are those which are highly polarizable and condensable and their adsorbed quantities depend on their boiling temperature. In physisorption, the quantity of the adsorbate is proportional to the specific surface of the substratum and independent from its shape. Physisorption also behaves differently according to the surfaces and the materials involved with the possibility of producing multiple layers of adsorbates. When it happens, the enthalpy of the bond in the adsorbate layers above the first one depends only on the interactions

Table 7.1. Maximum observed values for physisorption enthalpies

Adsorbates	$\Delta_{ad}H$ [kJ/mol]
CH_4	−21
H_2	−84
H_2O	−59
N_2	21

between the molecules of the adsorbate, consequently the enthalpy coincides with the latent heat of vaporization and therefore cannot form multiple layers (by condensation) in supercritical conditions, as in the case of hydrogen when is at a temperature above 33 K.

If the bond possesses a specific nature, like a covalent bond, the adsorption is called *chemisorption*. This type of adsorption is also an exothermic process. It is slower than physisorption and has an enthalpy equal to the enthalpies of the bond formation (200–400 kJ/mol). The adsorbed molecules tend to maximize the coordination number[3] with the adsorbent and the length of their bonds is usually shorter than that in a physisorption. Chemically, the adsorbed molecules can be broken and remain separated on the surface of the adsorbent, making the surface a potential catalyst for chemical reactions.

Desorption is the opposite phenomenon in which the adsorbed molecules are released.

7.3.2 Empirical Models of Molecular Interactions

Since there are no theoretical models to describe the interatomic and intermolecular interactions in the phenomenon of adsorption, empirical models have been developed for the purpose. One of the most commonly-used is the *Lennard-Jones potential*.

The formulation of the model considers both the Van der Waals' force at a long distance and the repulsion forces that occur at a very short range, which are caused by the interactions between the electronic distribution of different atoms or molecules and by the *Pauli exclusion principle*[4]. The potential, also known as the *12-6 potential*, is expressed as:

$$V\left(r\right) = 4\varepsilon \left[\left(\frac{\sigma}{r}\right)^{12} - \left(\frac{\sigma}{r}\right)^{6} \right] \qquad (7.20)$$

where ε is the depth of the potential well and σ is the collision diameter of the molecules determined by the kinetic theory of gases. The Van der Waals' forces at

---

[3] Coordination number means the number of molecules and ions linked to a central atom in a structure. In crystallography the term is used to indicate the number of atoms directly adjacent to a single atom in a definite crystalline structure.

[4] The *Pauli exclusion principle* states that no two identical fermions can have the same quantum numbers. A *fermion* is a particle that has half-integer spin and follows the *Fermi-Dirac statistics*. Protons, neutrons and electrons are examples of fermions.

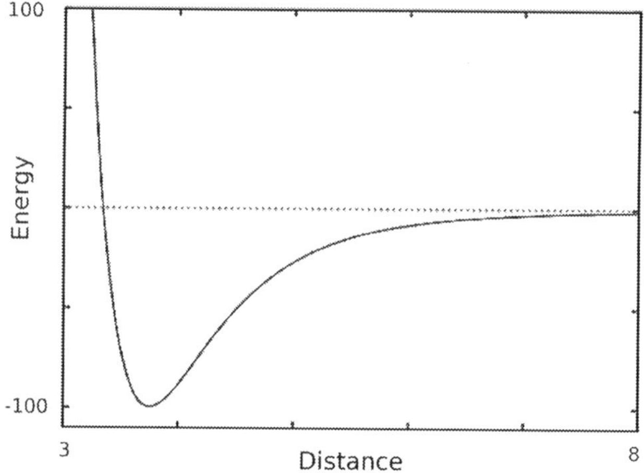

Fig. 7.1. Interaction energy of diatomic argon

long ranges are described by the r^{-6} term while the repulsion forces at short ranges are expressed by the r^{-12} term. Figure 7.1 shows the interaction energy for diatomic argon.

The *Morse function* is used to describe the potential energy in the interactions among diatomic molecules:

$$V_{Morse}(R) = D_e \ \{1 - \exp[-\beta(R - R_e)]\}^2 \qquad (7.21)$$

where D_e is the molecule dissociation energy, β is a constant and R_e is the atomic distance when the potential energy is at its minimum.

Similarly to the Lennard-Jones potential, the Morse function describes very accurately the molecular dissociation at long ranges, the repulsion action at short ranges and the spatial region of stability where the potential energy acquires the minimum value. The function has a value of zero when $R = R_e$, the point of minimum of the potential energy. When $R >> R_e$, the function value becomes $V = D_e$, representing the diatomic molecule dissociation energy at infinite distance. When $R << R_e$, the Morse function becomes:

$$V_{Morse}(R) = D_e \ [1 - \exp(\beta R_e)]^2 \qquad (7.22)$$

and corresponds to the positive and strong repulsion force among the atoms at short ranges.

It is possible to determine the type of adsorption according to intersection points of the Morse function curves, which represent the interaction between the adsorbate and the adsorbent. For example, in Figure 7.2 (a), when the molecule approaches the adsorbent, it can enter in a precursor state of physisorption and eventually trigger dissociation with chemisorption. In Figure 7.2 (b), chemisorption can occur only if the molecule has a greater energy than the activation energy E_a. When the activation threshold is surpassed and chemisorption occurs, the reaction will generally be

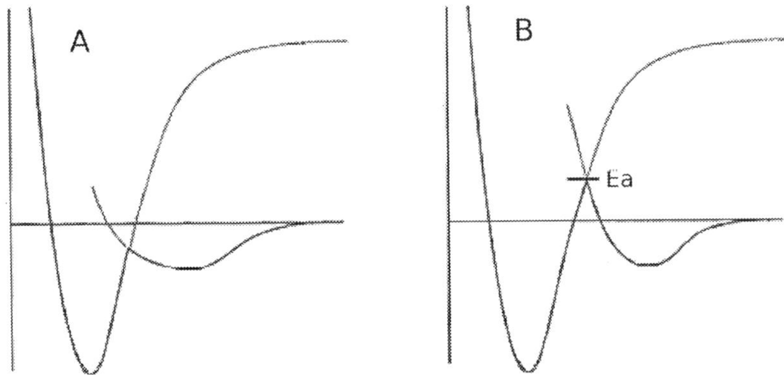

Fig. 7.2. Adsorption profiles of a diatomic molecule

slow and will require an increase in temperature for the activation of the adsorbent-adsorbate couple.

7.3.3 Adsorption and Desorption Velocities

The adsorption velocity is given by:

$$v_a = -\frac{dp}{dt} = k_a A p \tag{7.23}$$

where t is the time, p is the pressure, A is the surface available for adsorption reaction and k_a is the velocity coefficient of the adsorption in m^2/s.

Integrating Equation (7.23) yields:

$$p = p_0 \exp(-k_a A t) \tag{7.24}$$

which makes it possible to obtain the adsorption velocity coefficient if the adsorbent surface is known and if the desorption kinetics are negligible with respect to that of adsorption.

From the kinetic theory of gases, the number of the molecules colliding a surface per unit area is given by:

$$J_N = \frac{1}{4}\frac{N}{V}\bar{v} \tag{7.25}$$

where $\bar{v}$ is the average velocity of the molecules and N/V the molecule number per unit volume.

The probability that a molecule gets adsorbed onto a surface is called *sticking probability*. It is given by:

$$s = \frac{k_a}{J_N}. \tag{7.26}$$

In experiments, the adsorption velocity is measured by placing the adsorbent in a container which is then heated and degasified under vacuum. A given quantity of adsorbate gas is then pressurised into the container. By measuring the pressure decrease over time, it is possible to obtain information on the adsorption kinetics.

The desorption velocity is calculated as:

$$v_d = \frac{dN}{dt} = k_d N \tag{7.27}$$

where N is the number of molecules adsorbed onto the surface and k_d is expressed by the *Arrhenius' law*[5]:

$$k_d = A \exp\left(-\frac{E_a}{RT}\right) \tag{7.28}$$

where A is a pre-exponential factor independent from temperature and E_a is the activation energy.

The desorption velocity is obtained by heating the adsorbate-adsorbent complex and by measuring the change in pressure of the desorbed gas. With the increase of temperature, the adsorbate is released faster but the number of the released molecules decreases gradually, resulting in a peak in the *p-T* characteristic curve of desorption. The following relation is valid for the peak point of the desorption curve:

$$\frac{dv_d}{dT} = 0 = \frac{dk_d}{dT} N + \frac{dN}{dT} k_d. \tag{7.29}$$

Replacing k_d from the Arrhenius' law in the first addendum in Equation (7.29) the result becomes:

$$\frac{dk_d}{dT} = \frac{E_a}{RT^2} k_d. \tag{7.30}$$

By imposing a heating process according to the linear relation[6]:

$$T = T_0 + \alpha t. \tag{7.31}$$

By substitution in the second addendum in Equation (7.29):

$$\frac{dN}{dT} = -\frac{dN}{dt}\frac{dt}{dT} = -\frac{k_d N}{\alpha}. \tag{7.32}$$

Finally, from Equation (7.30) and (7.32) in Equation (7.29):

$$\frac{T_m^2}{\alpha} = \frac{E_a}{R k_d} \tag{7.33}$$

which provides the value of the temperature of the maximum desorption.

After extraction of the natural logarithm:

$$\ln\frac{T_m^2}{\alpha} = \ln\frac{E_a}{R} - \ln k_d = \ln\frac{E_a}{RA} - \frac{E_a}{RT}. \tag{7.34}$$

---

[5] The *Arrhenius' law* is an empirical formula that relates the rate of a chemical reaction (or, equivalently, its reaction constant) to temperature.
[6] α typically is around 20 K/s.

By plotting $(\ln T_m^2/\alpha)$ with α as the independent variable as a function of $1/T$, the result is a straight line whose gradient and coordinates of origin indicate the activation energy and the pre-exponential factor A respectively.

In some cases there can be more than one peak in the desorption curve, such as when the adsorption occurs on different crystalline surfaces.

7.3.4 Experimental Measurements of Adsorption and Desorption

In laboratory conditions, a common method to evaluate the quantity of the adsorbate on an adsorbent surface is to inject a known quantity of gas in a vessel containing the adsorbent and then to calculate the number of gas molecules adsorbed by measuring the change in gas pressure. Another technique is to measure the difference between the input and output flows of the gas after it passes through the container of the adsorbent.

In *flash desorption*, the adsorbent-adsorbate complex is heated quickly and the resulting increase in pressure grants the measurement of the quantity of the desorbed gas.

Gravimetric measures can be taken by using a *quartz crystal microbalance*: the differences in the oscillation frequencies between the adsorbent and the adsorbent-adsorbate complex, as recorded by the piezoelectric sensors of the device, correlate to the quantity of the molecules that have been adsorbed.

7.3.5 Adsorption Isotherms

The *Fractional coverage* θ is defined as the ratio between the number of the sites occupied by the adsorbate and the number of the sites available for adsorption.

Equivalently, it can be defined as:

$$\theta = \frac{V}{V_\infty} \tag{7.35}$$

where V is the volume of the adsorbate and V_∞ is the volume of the adsorbate corresponding to the complete adsorption of a layer of adsorbate. The *rate of adsorption* $d\theta/dt$ is the change of the fractional coverage over time.

The variation in the fractional coverage as a function of the pressure at a constant temperature is called *adsorption isotherm*. There are different formulations based on different hypotheses to describe how the quantity of the adsorbate changes according to pressure. One classic formulation is the *Langmuir's isotherm*, which is based on the following assumptions:

- adsorption is mono-layer and there are no other overlapping layers of adsorbed molecules,
- all adsorption sites have the same probability of being occupied by adsorbates;
- the surface of the adsorbent is perfectly uniform;
- the probability of a molecule being adsorbed in a site is independent from whether the adjacent spaces have already been occupied by other molecules.

The adsorption velocity based on these hypotheses is determined by the partial pressure of the gas p and the remaining adsorption sites available $N(1-\theta)$, as expressed in this relation:

$$v_a = \frac{d\theta}{dt} = k_a\,pN\,(1-\theta).$$

(7.36)

The desorption velocity is:

$$v_d = \frac{d\theta}{dt} = -k_d\,N\,\theta.$$

(7.37)

The two velocities are the same when the balance is reached and therefore the Langmuir's isotherm is expressed as:

$$\theta = \frac{Kp}{1+Kp} \quad \text{with} \quad K = \frac{k_a}{k_d}.$$

(7.38)

The isotherm curves in Figure 7.3 show how the fractional coverage increases with pressure. The saturation value reaches 1 only when the pressures are very high, which is coherent to the fact that the gas molecules are driven to occupy every remaining site available. Different temperatures produce different curves and the value of K also varies with the temperature, modifying the ratio between k_a and k_d. As it can bee seen in Figure 7.3, for a reference pressure value, higher values of K provide higher fractional coverages and different adsorption isotherms.

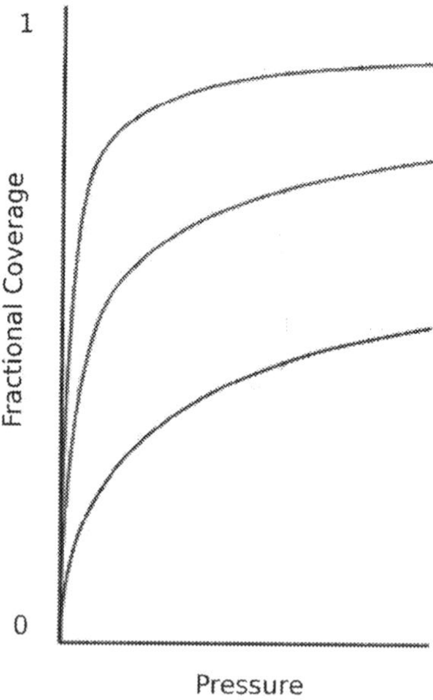

Fig. 7.3. Langmuir's isotherms for physisorption

7.3.6 Thermodynamics of Adsorption

Normally adsorption is a reaction with $\Delta G < 0$ and an entropy variation $\Delta S < 0$ since tthe mobility of the adsorbed molecules is reduced. For this reason, from the equation:

$$\Delta G = \Delta H - T\Delta S < 0 \tag{7.39}$$

it can be deduced that $\Delta H < 0$, with adsorption being therefore an exothermic process[7].

The adsorption enthalpy is determined by the adsorbent surfaces and by the number of occupied adsorption sites. If the adsorbed molecules tend to repel one another (i.e. carbon monoxide from palladium), the process becomes less exothermic with the increase of fractional coverage. If, instead, the molecules are more inclined to attract to each other (i.e. oxygen on tungsten), then the adsorption tends to occur in an island-like pattern with adsorption probabilities that are higher along the borders than in other sites. If the internal energy increases, the phenomenon of *order-disorder transitions* can occur when the thermal motion prevails over attraction forces between the molecules.

In general, the enthalpy of adsorption behaves in the exact opposite way of that of θ: the adsorption enthalpy decreases if the number of occupied sites increases. Contrary to one of the hypotheses of Langmuir's isotherm, the possibilities of the adsorption sites being occupied or not are not energetically equiprobable; instead, those with higher bonding energy have higher probability of being occupied.

The adsorption enthalpy calculated with constant θ is called *isosteric enthalpy*, given by the following equation:

$$\frac{\Delta H_{ads}}{RT^2} = -\left(\frac{\partial \ln p}{\partial T}\right)_\theta \tag{7.40}$$

which is similar to the *Clausius-Clapeyron's equation* but with the opposite enthalpy sign, since it is a condensation and not an evaporation process.

In case of a Langmuir's isotherm, it can be expressed as:

$$\frac{\theta}{1-\theta} = K p \tag{7.41}$$

and with θ being constant, it becomes:

$$d\ln k + d\ln p = 0. \tag{7.42}$$

The isosteric enthalpy in this case is given by:

$$\frac{\Delta H_{ads}}{RT^2} = \left(\frac{\partial \ln K}{\partial T}\right)_\theta. \tag{7.43}$$

It can be noted how the equation is similar to the *Van't Hoff's equation*[8] when ΔH_{ads} is replaced by ΔH^0 at the equilibrium.

---

[7] Exceptions can exist for the adsorption exothermicity, particularly when an adsorbed molecule disassociates and obtains a high translational freedom on the adsorbent lattice.

[8] The Van't Hoff's equation is a linear expression of the variation of the equilibrium constant of a chemical reaction according to the change in temperature.

7.3.7 Other Isotherms

The hypotheses on which the Langmuir's isotherm is based can cause significant divergences from the experiment results. For example, the assumption on the site equiprobability of adsorption contrasts with the adsorption enthalpy measures that show how such enthalpy declines when the functional coverage increases. For this reason, other isotherms based on less restrictive hypotheses have been introduced.

The *Temkin's isotherm* is expressed as:

$$\theta = c_1 \ln(c_2\, p) \tag{7.44}$$

with c_1 and c_2 as the constants determined by experiments. This isotherm assumes that the enthalpy varies linearly with the pressure.

The *Freundlich's isotherm*, is given by:

$$\theta = c_1\, p^{\frac{1}{c_2}} \tag{7.45}$$

which assumes instead that the enthalpy varies logarithmically with pressure.

If the adsorption occurs with condensation above the first adsorbate layer, the commonly used isotherm in this case is the one developed by Brunauer, Emmett and Teller (*BET isotherm*):

$$\frac{V}{V_{mon}} = \frac{c\,z}{(1-z)\,[1-(1-c)\,z]} \tag{7.46}$$

where

$$c = \exp\left(\frac{\Delta_{des}H^0 - \Delta_{vap}H^0}{R\,T}\right). \tag{7.47}$$

V_{mon} is the volume of the adsorbate considering only the first stratum, $z = \frac{p}{p^*}$ and p^* is the vapour pressure of a layer of adsorbate thicker than a molecule. Due to the fact that there can be more than one stratum of adsorbate, the curve does not saturate like a mono-layer isotherm, but grows indefinitely instead.

7.3.8 Classification of Isotherms

Apart from a few exceptions, the adsorption isotherms can be categorized according to the Brunauer's classification [7] into five types.

The isotherms of type I describe the typical mono-layer adsorption of the Langmuir's isotherm. Usually chemisorption is a mono-layer phenomenon and follows the curve of type I.

Type II and III correspond to multilayer adsorptions. The fractional coverage of type II increases rapidly at the beginning, then tends to saturate within a range of pressure values, before rising again with a nearly exponential trend. Type III instead manifests an exponential tendency for all pressure values.

Type IV and V describe adsorptions on porous sub-strata. Type IV isotherm initially shows a curve similar to type II, but after a certain pressure value it saturates to a limited coverage value. Type V starts with an exponential trend but ends up saturating in a way similar to type IV isotherms.

If the dimension of the pores is around 10 nm, the desorption curves can be different from those of adsorption, presenting a hysteresis caused by the difference of pressure when the adsorbates condenses in the pores with respect to the pressure at desorption. The explanation of this phenomenon is given by the *Kelvin's equation*, under which a pressure gradient exists and is normal to a curved surface, as per:

$$\ln \frac{p}{p_0} = \frac{2\gamma V_m}{r\,\mathrm{R}\,T} \tag{7.48}$$

where p_0 is the saturated vapour pressure, γ is the surface tension, V_m is the molar volume and r is the radius of the curve of the surface.

7.3.9 Carbon Materials for the Physisorption of Hydrogen

7.3.9.1 Nanotubes

Nanotubes are cylindrical tubes of graphite with a diameter of a few atoms and a length of tens of thousands of atoms.

At certain temperatures and pressures, carbon nanotubes have the ability to bind a hydrogen atom to a carbon atom in the lattice. When the temperature rises, the thermal motion releases the hydrogen atoms which return to the gaseous phase and can be taken out from the storage tank for further uses.

In particular situations, carbon atoms can form a spherical structure called *fullerene*, which, after further relaxation, can roll and form the typical cylindrical structure of carbon nanotubes. Nanotubes, similar to fullerenes, can be seen as one of the many allotropes of carbon.

The different types of nanotubes can be divided into the following two categories:

- *Single-Wall Carbon Nanotube* (SWCNT), composed only by one layer of graphite atoms in the shape of a cylinder;
- *Multi-Wall Carbon Nanotube* (MWNT), consisting of multiple layers of graphite rolled in concentric cylinders.

The main body of the nanotube is formed mainly by hexagons while the enclosing parts are made by both hexagons and pentagons. Irregularities in the pentagon-hexagon configuration can cause structure defects and imperfections which can deform the cylinder. The diameter of a tube ranges from a minimum of 0.7 nm to a maximum of 10 nm. The very high ratio between the length and the diameter means that these tubes can be considered as one-dimensional structures.

A single-wall nanotube is very resistant to traction. It possesses some interesting electric properties: depending on its diameter and its chirality[9] (caused by the way the carbon-carbon bonds are formed one after the other along the tube circumference), it can be either a conductor like metal or a semiconductor like silicon.

---

[9] A molecule is *chiral* if it does not have an internal plane of symmetry, hence does not have a super-imposable mirror image.

With a wide surface area, which is estimated up to 1350 m^2/g per face, nanotubes can also be used for gas storage. To evaluate the maximum capacity, the numerical simulations calculate both the internal and external surfaces of the nanotube, as well as the inter-tube area. The maximum percentage of adsorption for a nanotube is calculated to be around 10%, but this only refers to neat and perfectly aligned nanotubes. In reality they can be unaligned with many points of contact that reduce the number of available sites for the interaction with the gases.

7.3.9.2 Activated Carbons

Activated carbon (AC) is a type of carbon whose structure is highly porous and therefore is characterised by a large *specific surface area*[10]. Compared to carbon nanotubes, activated carbon is simpler to produce and has a greater specific surface area than any other well-known carbon structure: for instance, the commercially available activated carbon AX-21 has a specific surface area up to 2800 m^2/g.

Activated carbons can be produced by physical or chemical activation. The physical activation processes raw materials like wood or carbon in two phases:

- carbonization (or pyrolysis) carried out at high temperatures without oxygen;
- oxidation which exposes the carbonized materials to highly oxidizing materials, such as water steam, at temperatures from 800 to 1100 °C.

The diameter of the pore of the structures obtained with this process is less than 50 nm, qualifying the activated carbons as a type of micro or meso-porous materials. Chemical activation is more often used in processing peat and other wooden materials. The procedure involves exposing the materials to various kinds of chemicals at high temperatures, such as phosphoric acid (H_3PO_4) or a salt like zinc chloride ($ZnCl_2$). The materials are dehydrated by these chemicals and become a paste, which subsequently goes through a slow carbonization at temperatures from 500 to 800 °C and then is cleansed and rinsed with distilled water. Compared to physical activation, this procedure is faster and cheaper, even though it tends to pollute the final product. The pore diameter of the AC produced by this process is longer than 50 nm, making the carbon a macro-porous material.

Activated carbons with smaller pores are more suitable for filtering fluids, while macro-porous activated carbons are better for processing fumes as they do not cause flow rate reductions. As for the treatment of aeriforms, activated carbons can be used in air conditioning systems, cigarette filters, industrial CO_2 filtration plants and incinerators to purify the smoke produced by the burning of waste. Activated carbons can also be mixed with foam or fibre to produce different kinds of materials. They are also used in industrial waste water treatment, groundwater filtration and drinking water purification to remove substances like dioxins, heavy metals and hydrocarbons.

As already mentioned, the specific surface area characterises activated carbons in their applications. It is calculated by the *BET method*, named after its inventors Brunauer, Emmett and Teller. The method evaluates the *BET surface* which gives the

---

[10] The *specific surface area* is the total surface area per unit mass.

most important information to evaluate the performance of carbon nano structures. The value usually falls between 500 to 1500 m^2/g but can also reach up to 3000 m^2/g.

Another important parameter is the *ash residues* formed when the carbon is heated at 954 °C for 3 h in a porcelain container. These ashes take up from 3 to 10% of the total initial mass of the raw material. Hydrochloric acid can be used to rinse the carbon from ashes.

It is also essential to know the *toughness* of the raw material used because it indicates the length of service life of the AC filters. It depends on the nature of the raw materials and the type of treatment given.

Another significant criterion from the economic point of view is the material density. High density allows the production of small and durable filters but also reduces flow rates and applicable pressure ranges. The material density can refer to *bulk density* if it considers the structure of the material together with the volume of the pores, or to *skeletal density* if it concerns only the volume of the carbon material. In the case of AX-21, for example, the structural density is 0.3 g/cm^3 for powder materials and 0.72 g/cm^3 for AC pressed in pellets. Its skeletal density is 2.3 g/cm^3.

The specific measurements of the porosity in activated carbon can be helpful in determining the adsorption features of the carbon structure. For example, the calculation of the *iodine number* determines the quantity of iodine adsorbed (measured in mg/g). Iodine structure consists of molecules around 10 angstrom and can therefore reveal the presence of the pores of the same dimension. The activated carbons used for water treatment generally have an iodine number ranging from 600 to 1200 mg/g.

Another similar method is to use the *methylene blue number*, which indicates the number of mg of methylene blue adsorbed in 1 g of AC. The dimension of methylene blue molecules is around 15 angstrom and they can reveal the presence of the pores of the same size. Other methods include the *molasses number*, which calculates the pores of 28 angstrom and the *carbon tetrachloride number*, which indicate the adsorption of tetra-chloromethane steam.

Activated carbons can be used to store hydrogen by means of physisorption, as will be detailed in Chapter 9.

7.3.10 Alternatives to Carbon Physisorption

At the moment researches are being conducted to find alternatives to physisorption storage by carbon: boron oxide (B_2O_3), for example, has demonstrated a very good adsorption capacity at a temperature of 115 K.

7.3.11 Zeolites

Zeolites are minerals with a regular and microporous crystalline structure. Because of their symmetrical molecular composition, they can be used as a molecular sieve with a higher selectivity than other minerals like silica and activated carbon, whose pore structures and dimensions are irregular.

The cations contained in the structure of zeolites can activate the procedure of *ion exchange*, which is the exchange of these cations with compatible ions contained in solutions and gases (such as hydrogen). When zeolites are heated, the ions can pass through to the internal structure and remain sequestered until the next heating.

There are about 50 different kinds of natural zeolites with various chemical properties and crystalline structures, but none of them has a gravimetric density higher than 2–3% by hydrogen weight. This performance is far too lower than the standard established by DOE and therefore recent researches have been concentrating on developing artificial zeolites with better storage capacity.

7.3.12 Metallic Hydrides

The earliest study performed on metallic hydrides dates back to 1866 when Graham first observed the phenomenon of adsorption of hydrogen by palladium. Hydrogen accumulates in a crystalline structure of a certain type of metal and forms a *metallic hydride*. The adsorption of hydrogen is facilitated by removing heat and by adding high pressure, while the increase of heat helps restore the embedded hydrogen to its former state.

This process is reversible and develops in two phases:

- *Hydrogenation or exothermic adsorption*: hydrogen is injected in the reaction device containing the metallic hydride to diffuse and to become attached to the hydride lattice. This process occurs at 3–6 MPa with the release of heat around 15 MJ/kg.
- *Dehydrogenation, or endothermic extraction of hydrogen*: by increasing the temperature to above 500 °C, hydrogen can be restored to its previous bi-atomic structure and released for re-utilisation.

Metallic hydrides are categorized according to their temperatures of hydrogenation: those with a hydrogenation temperature above 300 °C, like magnesium alloys, are better for storage purposes with good capacities. Current research is working on how to develop hydrides able to function at lower temperatures (under 100 °C) so that they can be used in combination with PEM fuel cells.

Given a certain isotherm, the charging process can be divided into three phases. The first phase (α) is the hydrogen diffusion in the metal structure. It is expressed by this relation:

$$\sqrt{p} = K_S x \tag{7.49}$$

where K_S is the Sievert constant.

In the second phase ($\alpha + \beta$) hydrogen begins to react with the metal and, regardless of the increase of the concentration of the gas, the pressure remains the same. The final phase (β) occurs when the pressure starts to climb while the concentration of the gas continues to increase. The equation describing the second and the third phases is:

$$\ln(p) = \frac{a}{T} + b \tag{7.50}$$

with:

$$a = \frac{\Delta H^{\alpha \to \beta}}{x R} \tag{7.51}$$

where b is a parameter determined by experiments.

The process is reversible and can be expressed as the following for a given metal M:

$$M + \frac{x}{2} H_2 \leftrightarrow MH_x + \Delta H^{\alpha \to \beta} \qquad (7.52)$$

where M stands for the metal, MH is the metallic hydride, x is the ratio between the quantities of hydrogen atoms and the metal atoms and $\Delta H^{\alpha \to \beta}$ is the enthalpy of formation of the metallic hydride.

The concentration of the hydrogen in a metallic hydride at equilibrium changes according to the temperature and the pressure of the gas: when the pressure rises and the temperature remains steady, the hydrogen concentration increases. It is therefore possible to determine the isotherm curves of the concentration variations according to the pressure and the gas temperature.

Hydrogen equilibrium pressure p_{eq} can be calculated also by using the Van't Hoff's equation:

$$\ln(p_{eq}) = \frac{\Delta H}{R T} - \frac{\Delta S}{R} \qquad (7.53)$$

where ΔH is the variation of the enthalpy of formation (in J/mol) and ΔS is the variation of the entropy of formation in (J/mol).

Using the method of thermal mass capacity, the temperature of the hydride layer T varies according to time. The heat needed for hydrogen charging and discharging can be computed as:

$$C_{mh} \frac{dT(t)}{dt} = Q_{mh}(t) - k\left[T(t) - T_a\right] \qquad (7.54)$$

where C_{mh} is the thermal capacity of the hydride, Q_{mh} is the heat yielded or absorbed by the hydride, which is released during hydrogen adsorption and required for desorption; k is the heat loss coefficient and T_a is the environment temperature.

When integrated into a fuel cell system, it is possible to retrieve part of the thermal energy produced during the cell operation for the thermal cycles of the hydride storage system.

7.4 Chemical Storage

7.4.1 Chemical Hydrides

Chemical hydrides are formed by a reversible reaction of hydrogenation in liquid mixtures at ambient pressures and temperatures For the purpose of hydrogen storage, they are capable of reaching good to high capacities. For example, lithium borohydride (LiBH$_4$) contains 18.5% hydrogen by mass with a reversibility of 13.8%. Not all the hydrogen can be released because part of it remains bound to lithium as per the following reaction:

$$LiBH_4 \leftrightarrow LiH + B + \frac{3}{2} H_2. \qquad (7.55)$$

LiH cannot decompose further as it is very stable; in fact its decomposition temperature (573 K) is higher than the fusion temperature of $LiBH_4$ (550 K).

Using sodium borohydride instead of lithium in a reaction with water and adopting ruthenium as the catalyst, the following reaction is obtained:

$$NaBH_4 + 2\,H_2O \rightarrow 4H_2 + NaBO_2 + 300\,kJ \tag{7.56}$$

with an obtainable gravimetric density around 7.5%. The hydrogen produced in this way is ideal to be used in PEM fuel cells as it can reach a very high degree of purity.

Alternatively, hydrogen can be obtained by directly decomposing sodium borohydride ($NaBH_4$) as per the following reaction:

$$NaBH_4 + 8\,OH^- \rightarrow NaBO_2 + 6\,H_2O + 8\,e^-. \tag{7.57}$$

The sodium borate can be further retrieved to regenerate sodium borohydride.

An important advantage of this type of technology is its long length of storage capacity, which extends for more than 100 days. The operating costs are also comparable to those of traditional fossil fuels. It also has the potential to employ the existing infrastructure for safe and cheap transport and distribution. Sodium borohydride has already found applications in the aerospace and the automotive industries.

References

1. Aranovich G L, Donohue M D (1996) Adsorption of supercritical fluids. Journal of colloid and interface science 180:537–541
2. Atkins P, De Paula J (2006) Atkins's Physical Chemistry. Oxford University Press, Oxford
3. Aziz R A (1993) A highly accurate interatomic potential for argon. J. Chem. Phys. 6 (99):4518
4. Bénard P, Chahine R (2001) Determination of the adsorption isotherms of hydrogen on activated carbons above the critical temperature of the adsorbate over wide temperature and pressure ranges. Langmuir 17:1950–1955
5. Bénard P, Chahine R, Chandonia P A, Cossement D et al (2007) Comparison of hydrogen adsorption on nanoporous materials. J. Alloys and Compounds 446–447:380–384
6. Bénard P, Chahine R (2007) Storage of hydrogen by physisorption on carbon and nanostructured materials. Scripta Materialia 56:803–808
7. Brunauer S, Emmett P H, Teller E (1938) Adsorption of gases in multimolecular layers. J. Am. Chem. Soc. (60):309–319
8. Cheng J Yuan X, Zhao L, Huang D et al (2004) GCMC simulation of hydrogen physisorption on carbon nanotubes and nanotube arrays. Carbon 42:2019–2024
9. Chen X, Zhang Y, Gao X P, Pan G L et al (2004) Electrochemical hydrogen storage of carbon nanotubes and carbon nanofibers. International Journal of Hydrogen Energy 29:743–748
10. Frackowiak E, Beguin F (2002) Electrochemical storage of energy in carbon nanotubes and nanostructured carbons. Carbon 40:1775–1787
11. Fukai Y (1993) The Metal-Hydrogen Systems: Basic Bulk Properties. Springer-Verlag, Berlin

12. Gregg S J, Sing K S W (1982) Adsorption, Surface Area and Porosity, 2nd ed. Academic Press, New York

13. Hirscher M, Becher M, Haluska M, Quintel A et al (2002) Hydrogen storage in carbon nanostructures. Journal of Alloys and Compounds 330–332:654–658

14. Hirscher M, Becher M, Haluska M, von Zeppelin F et al (2003) Are carbon nanostructures an efficient hydrogen storage medium? Journal of Alloys and Compounds 356–357:433–437

15. Iijima S (1191) Helical microtubules of graphitic carbon. Nature 354:56–58

16. Jhi S-H, Kwon Y-K, Bradley K, Gabriel J-C P (2004) Hydrogen storage by physisorption: beyond carbon. Solid State Communications 129:769–773

17. Lamari Darkrim F, Malbrunot P, Tartaglia G P (2002) Review of hydrogen storage by adsorption in carbon nanotubes. Int. J. of Hydrogen Energy 27:193–202

18. Panella B, Hirscher M, Roth S (2005) Hydrogen adsorption in different carbon nanostructures. Carbon 43:2209–2214

19. Reilly J J (1977) Metal hydrides as hydrogen energy storage media and their application. In: Cox K E, Wiliamson K D (eds.) Hydrogen: Its Technology and Implications, vol. II. CRC Press, Cleveland, pp. 13–48

20. Sandrock G, Thomas G (2001) IEA/DOC/SNL on-line hydride databases. Appl. Phys. A72 153

21. Schimmel H G, Nijkamp G, Kearley G J, Rivera A et al (2004) Hydrogen adsorption in carbon nanostructures compared. Materials Science and Engineering B108:124–129

22. Thomas K M (2007) Hydrogen adsorption and storage on porous materials. Catalysis Today 120.389–398

23. Vanhanen J P, Lund P D, Hagström M T (1996) Feasibility study of a metal hydride hydrogen store for a self-sufficient solar hydrogen energy system. Int. J. Hydrogen Energy 21:213–221

8

Other Electricity Storage Technologies

Finding an efficient solution to store electric energy is an important goal to achieve for both small and large-scale applications. This chapter explores some of the best proposals coming from the research in the field, such as electrochemical batteries, compressed air, pumped water, pumped heat, super-capacitors and other innovative technologies.

8.1 Introduction

The following are some of the most promising technologies for electricity storage with the prospect of achieving substantial progress in the near future:

- electrochemical storage;
- ultra-capacitors;
- compressed air;
- underground pumped water;
- pumped heat;
- natural gas production;
- flywheels;
- superconducting magnetic energy storage.

8.2 Electrochemical Storage

A *battery*, or *accumulator*, is an electrochemical device capable of converting electric energy into chemical energy, storing it during the charging cycle and converting it back to electricity during the discharging cycle. Thanks to this feature, it is widely used in many different sectors.

The total energy $E(t)$ stored in a battery is calculated as:

$$E(t) = E_{in} + \int_{o}^{t} U_B(t) I_B(t) \, dt \tag{8.1}$$

Zini G., Tartarini P.: Solar Hydrogen Energy Systems. Science and Technology for the Hydrogen Economy. DOI 10.1007/978-88-470-1998-0_8, © Springer-Verlag Italia 2012

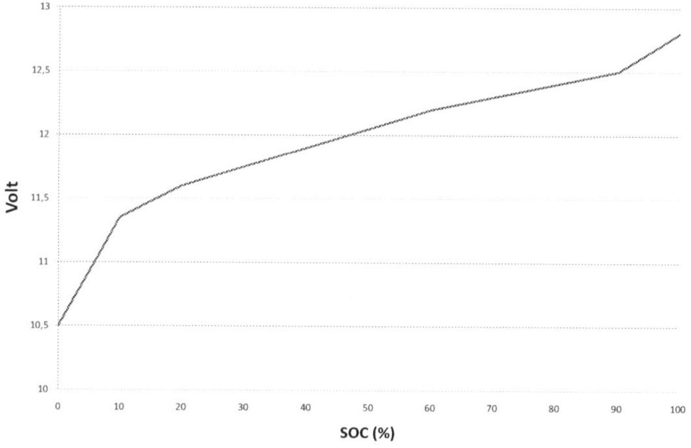

Fig. 8.1. Variation of the open-circuit voltage of a 12 V battery as a function of SOC

where E_{in} is the energy stored in the battery at initial conditions, U_B and I_B the voltage and current in the battery. Knowing the maximum energy E_{max} that the battery is capable of storing, which depends on its internal characteristics, the battery *State Of Charge* (SOC) is defined by:

$$SOC = \frac{E(t)}{E_{max}}. \tag{8.2}$$

The battery voltage is usually mantained within a well-defined range to avoid irreversible damages to battery. Therefore, a control system manages the charge and discharge cycles with a minimum SOC usually around 20–30% of E_{max} and a maximum SOC around 80–90% of E_{max}.

Figure 8.1 shows the variation of the open-circuit voltage of a 12 V electrochemical battery.

Conventionally, during charging the sign of the current I_B provided to the battery is positive, while it is negative during battery discharging.

The battery voltage $U_B(t)$ can be obtained from the following equation:

$$U_B(t) = (1 + \alpha t)\, U_{B,0} + R_i\, I_B(t) + K_i\, Q_R(t) \tag{8.3}$$

where t is time, α is the auto-discharge speed in (Hz), $U_{B,0}$ is the open-circuit voltage at $t = 0$, R_i is the battery internal resistance, K_i is a coefficient taking into account battery polarisation phenomena (in Ω/h) and $Q_R(t)$ is the accumulated charge (in Ah).

An important function of these batteries lies in their ability to smooth voltage fluctuations during operations when other devices in the energy system are being switched on and off so that the quality of service to the load can remain satisfactory.

Among the many different types of batteries available, the following three technologies will be covered here:

- valve regulated lead-acid batteries;
- lithium-ion batteries;
- vanadium redox flow batteries.

8.2.1 Valve Regulated Lead-Acid

The *Valve Regulated Lead-Acid* (VRLA) batteries are characterised by advantages such as no emissions during charging or normal operations and very few maintenance requirements. For this reason, they are very safe and suitable to use in hydrogen energy systems.

The two most commonly-used types of VRLA are:

- *Absorbent Glass Mat*: the electrolyte is absorbed in a fibreglass matrix granting higher purity levels since the electrolyte does not need to be capable of sustaining its own weight inside the battery vessel. Compared with traditional batteries, this device benefits from a lower internal resistance, endures higher operation temperatures, operates with lower discharge rates and possess higher power densities; therefore it is suitable to be used also in electric vehicles.
- *Gel Battery*: the electrolyte is in a gel form with the sulphuric acid electrolyte mixed with silica fume. This non-liquid feature allows the battery to avoid any problem of leakage or evaporation and the device does not need to be kept upright. The battery is also very resistant to external impacts, shocks, vibrations and high operation temperatures.

In VRLA batteries, the charge-discharge processes are generated by the movement of lead ions from one electrode to the other. The anode is made of lead while the cathode is made from porous lead-oxide.

During the discharge cycle, the anode lead is attacked by the sulphuric acid resulting in the formation of lead ions (Pb^{2+}) and the correspondent electrons (e^-) that flow in the external electric circuit with this oxidation reaction:

$$Pb \rightarrow Pb^{2+} + 2\,e^-. \tag{8.4}$$

At the anode, lead sulphate and hydrogen ions are formed following this reaction:

$$Pb^{2+} + HSO_4^- \rightarrow PbSO_4 + H^+ + 2\,e^-. \tag{8.5}$$

At the lead-oxide electrode, the electrons e^- from the external electric circuit react with Pb^{4+} with the reduction reaction:

$$Pb^{4+} + 2\,e^- \rightarrow Pb^{2+}. \tag{8.6}$$

At the cathode, lead sulphate and water are produced as per:

$$PbO_2 + HSO_4^- + 3\,H^+ + 2\,e^- \rightarrow PbSO_4 + 2\,H_2O. \tag{8.7}$$

The charging cycle is similar to discharging except that it occurs in the reversed direction and restores the initial electrolyte charge.

The output voltage is around 2 V as per the sum of the absolute values of the reaction potentials at the anode and at the cathode (-0.36 V and 1.69 V in standard conditions respectively). The batteries therefore need to be manufactured in stacks by connecting several single battery units in series, and the total number of the unit depends on the desired output voltage. Normally, for non stationary uses, the output voltage is 12 V, while for non-stationary applications the voltage can vary from 24 to 220 V.

The energy density is normally in the range of 20–40 Wh/kg.

8.2.2 Lithium-Ion

Lithium-ion rechargeable batteries (LIB) are commonly used in many different industrial, military and consumer applications. They benefit from high energy density, high reliability, good efficiency and large temperature operating ranges.

In conventional lithium-ion batteries, the electrolyte is a non-aqueous solution of lithium salts (like $LiPF_4$, $LiPF_6$, $LiAsF_6$, $LiClO_4$) in an organic solvent. The anode is usually in graphite, while the cathode can be made of metal oxides (i.e. $LiCoO_2$), crystal oxides as $LiMg_2O_4$ or poly-anions[1] like $LiFePO_4$. A thin sheet of special plastic separates the positive half-cell to the negative half-cell; the micro holes in the plastic leave the lithium ions to move from one half cell to the other.

During charging, the electric potential applied to the electrodes forces the lithium ions to migrate from the cathode (the positive electrode) to the anode (the negative electrode), where they embed in the porous cathode graphite material by *intercalation*[2]. During discharging, lithium ions take the reverse path through the separator diaphragm back to the cathode. Electric energy is therefore provided to the external circuit connected to the two electrodes.

The following is the reaction at the positive electrode (with x as the number of moles and the direction of charging going from the right to the left):

$$LiCoO_2 \leftrightarrow Li_{1-x}CoO_2 + xLi^+ + xe^-. \tag{8.8}$$

At the negative electrode, the inclusion process occurs as per:

$$xLi^+ + xe^- + 6C \leftrightarrow Li_x. \tag{8.9}$$

Normally, each cell yields a voltage of 3 to 4 V, which is higher than other types of batteries and makes this device lighter, more compact and more suitable for non-stationary applications. The energy density normally falls in the range of 80–180 Wh/kg.

Lithium though is a scarce resource, hence a large penetration of such batteries in the market could pose monopoly threats to the economy with increasing costs, political tensions and eventually the need to find substitutes in the medium term. Other disadvantages of the device include a short service life and the possibility of being damaged if a deep discharging occurs.

Despite these concerns, however, lithium-ion batteries will very likely become the forerunner to open up markets for electric vehicles and decentralised energy storage devices, before being replaced by new and more advanced energy storage materials.

---

[1] *Poly-anions* are molecules with negative charges positioned in different sites across the molecule structure.

[2] *Intercalation* is the inclusion of a different molecule into a pattern of other different molecules. The inclusion process is usually reversible.

8.2.3 Vanadium Redox

Vanadium is a transition metal with atomic number 23 and found in nature only in combined forms. The *vanadium redox* battery (VRB) employs vanadium in different oxidation states to store chemical potential energy. The VRB consists of a stack of cells where the reactions take place, two electrolyte tanks with solutions of vanadium salts in sulphuric acid and pumps to flow the electrolytes in the battery stacks (Figure 8.2). Stacks are separated by ionic membranes.

Vanadium is capable of forming solutions with four different oxidation states, which is a property shared only by other few elements in the periodic table. This is the reason why the VRB uses ions of the same metal in both the half-cells, thus eliminating the problem of electrolyte cross-contamination as can be observed in other types of redox batteries.

The electrons exchange at the electrodes according to the following reactions, in which the direction of charging is from the left to the right and the discharging goes in the opposite direction. At the positive electrode the reaction is:

$$VO^{2+} \leftrightarrow VO_2^+ + e^- \tag{8.10}$$

and at the negative electrode it becomes:

$$V^{3+} + e^- \leftrightarrow V_2^+. \tag{8.11}$$

The overall reaction is:

$$V^{3+} + VO^{2+} \leftrightarrow V_2^+ + VO_2^+. \tag{8.12}$$

During recharging, the VO^{2+} ions in the positive half-cell are converted to VO_2^+ ions and electrons enter the electric circuit in the anode of the battery. At the same

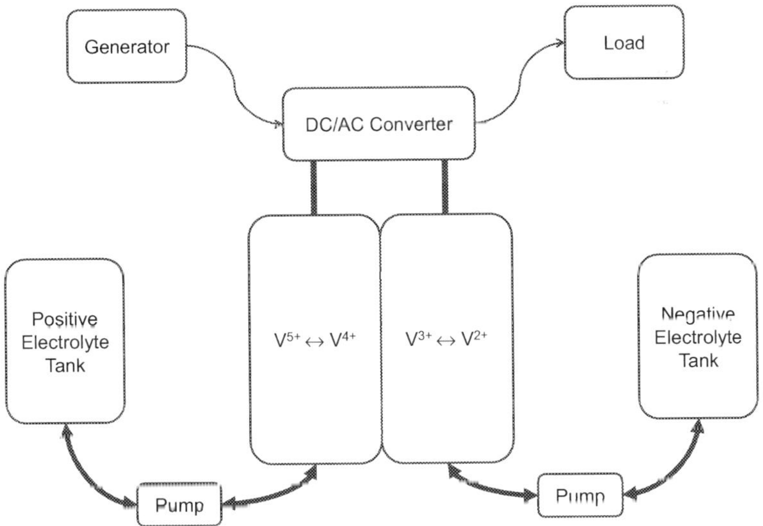

Fig. 8.2. Schematic of a vanadium redox battery

time, in the negative half-cell the electrons from the electric circuit react with the V^{3+} ions and convert them into V^{2+}. During discharging this process is reversed. The typical open-circuit voltage of one cell is 1.41 V at a temperature of 25 °C. The VRB full cycle efficiency is estimated to be around 70–80% with a power range of 0.5–100 MW and more than 10000 charging–discharging cycles when the electrolyte operates at 10–35 °C. Different from an electrochemical battery, the duty cycle can be as high as 100% as there is no need to avoid deep discharges. A VRB with a storage capacity of 40 kWh weight around 5000 kg with a footprint of 1.5 m^2 and a height of 2 m. The energy density is about 14 Wh/kg but can be reduced in half if all the satellite systems are included in the calculation.

The advantages of a VRB are manifold. First of all, the response time is very short, usually less than one millisecond at full charge. Also, the battery allows overloads of as much as 400% for ten seconds, and its capacity can be virtually unlimited by simply refilling the tanks. Furthermore, they show no memory effects whatsoever or display any sign of damage in case the two different electrolyte liquids mix with each other. However, the system is fairly complex and needs a large volume to achieve acceptable storage standard, thus resulting in its low energy-to-volume ratio. If a higher volumetric energy density can be achieved with future developments, the refill capacity of the VRB can help widen its non-stationary applications: for example, electric vehicles equipped with VRB in principle can be replenished at the fuel distribution stations just as quickly as traditional vehicles.

8.3 Ultra-capacitors

Capacitors are widely used in electronics applications like filters and control circuits. In their simplest form, they consist of a couple of electrically-conductive surfaces separated by a dielectric material. When a voltage is applied on the conductors, an electrostatic field is formed due to the positive charges appearing on one plate and the negative charges on the other. The energy that is consequently stored in the capacitor is given by:

$$E = \frac{1}{2} C U^2 \tag{8.13}$$

where U is the voltage applied to the capacitor and C is the *capacitance* defined as:

$$C = \varepsilon \frac{S}{d} \tag{8.14}$$

with d the distance between the two surfaces of area S and ε the dielectric constant of the insulator between the two electrodes. To increase the energy storage capacity, it is necessary to have a higher dielectric constant and a larger surface area with a shorter distance in between.

An *ultra-capacitor* is a type of electrochemical capacitor that stores the electric energy in the *Helmholtz double layer*[3] formed at the solid-electrolyte interface.

---

[3] A *double layer* is a structure of two parallel layers of electric charges formed on the surface of a solid immersed in a liquid. One layer is composed by ions adsorbed to the solid surface

The electric ionic charges in the electrolyte move to the electrode and form an electrostatic field with the electronic charges on the electrode surface. The amount of energy stored in the device then depends on the concentration of the electrolyte and on the physical characteristics of the ions. The electric field in the electrochemical double layer can reach very high values in the range of 10^6 V/cm.

An additional capacitance, called *differential capacitance*, is present in the *electric double layer* (EDL) formed on the electrode surface:

$$C_\Delta = \frac{d\sigma}{d\psi} \tag{8.15}$$

where σ is the surface charge and ψ is the electric potential on the electrode surface.

The electrodes can be made of metal oxides like RuO_2, IrO_2, polymeric materials or porous carbon with a surface area of 2500 m^2/g. A porous electrode indeed possesses a larger reaction surface and is able to create higher capacitances and to improve energy storage capacity. The electrolyte can be organic or in the form of a solution in water of acids like H_2SO_4. The combination of virtually unlimited charging-discharging cycles with the wide ranges of specific power ($10-10^6$ W/kg) and specific energy (0.05–10 Wh/kg) makes these devices a prominent candidate for many energy storage applications.

8.4 Compressed Air

The technology of *compressed air energy storage* (CAES) has been in development from the '60s, but only a few plants have been installed and put in operation ever since. The earlier systems required a complex underground compressed air storage, but recent technological advances have made above-ground systems a viable option.

An isothermal CAES has been recently developed with the design of two containers and an electrically-operated motor/generator. When energy needs to be stored, the motor pump compresses air from one container to the other and the air is simultaneously cooled by the spraying of water to keep a constant air temperature. The heated water is then stored in a reservoir. When the procedure is reversed, the air expands by passing through the pump which works as an electric energy generator at this point. The previously heated water is sprayed back to prevent the air from cooling and the air is then stored in the other container. This isothermal compression cycle achieves a higher efficiency than non-isothermal ones and further simplifies the design and the construction of the containers and the piping.

---

by chemical reactions, the other is formed due to the electrostatic interactions between the adsorbed ions with the ions in the liquid.

8.5 Underground Pumped Water

Hydroelectric plants pump upstream water during off-peak hours to take advantage of higher prices during peak hours. This practice is not meant to store energy, but rather to arbitrate on energy prices. Since hydroelectric plants are already fully exploited, it is difficult to think of them as an additional storage system. This is even more true with new hydroelectric plants since they are often more difficult to construct due to their environmental impacts and the scarcity of new water sources and reservoirs.

As an alternative to these plants, it is possible to drill a pair of parallel boreholes in the ground as deep as hundreds of meters: the main borehole acts as the deep storage pipe while the smaller one functions as the return pipe. Both holes are connected to a pump turbine. A weight stack is placed in the main pipe and is free to move up and down.

During storage, the turbine uses surplus energy to pump water through the return pipe to the bottom of the storage pipe; the weight stack is hence forced to move up, storing energy in the form of potential energy. During production, the stack descends on its own weight while water is pushed back to the return pipe. The potential energy stored in the weight stack height is converted to electric energy in the turbine. The conversion efficienty is measured to be as high as 80%. An area of 7-10 acres can host a storage capacity around 2 GW.

8.6 Pumped Heat

This is a method similar to section 8.5 except that it adopts heat instead of water. In this case, the reservoirs are two containers filled with gravel which are connected to a compressed gas pipe and a heat pump. The pump employs the surplus energy to heat the compressed gas up to 500 °C. The gas is then sent to one container to exchange heat with the gravel.

When energy is needed, the cycle is reversed and the gas cooled down to -160 °C by expansion. The gas exchanges heat with a second gravel container, with the heat pump working in reverse and converting thermal energy back to electricity.

A 16 MWh system would cover a surface of 128 m^2 with an overall efficiency estimated at 70–80%.

8.7 Natural Gas Production

Electricity can also be stored in the form of natural gas by a two-step process: first the surplus energy is applied in a usual electrolysis process to separate water into oxygen and hydrogen. The hydrogen is then employed to react with CO_2 to yield methane as the product.

The efficiency is estimated at 60%. Once carbon sequestration becomes a widespread practice, natural gas storage can be an intelligent solution to exploit carbon dioxide which would otherwise be stored unproductively in underground caves or containers.

8.8 Flywheels

The ability of spinning objects like *flywheels* to store kinetic energy has been known by mankind since the neolithic age. Flywheels are a widely-used mechanical device as well as a mechanical energy storage system that can replace traditional electrochemical batteries in some cases.

A flywheel is usually a spoked disk or cylinder capable of storing a significant amount of rotational energy. When shafted to an electric motor/generator, the flywheel can store energy when the motor increases its angular speed, or release energy to the generator when the latter acts as a brake to decrease the flywheel angular speed.

The energy of a spinning flywheel is given by:

$$E = \frac{1}{2} I \omega^2 \tag{8.16}$$

where ω is the angular velocity and I is the inertia momentum, which is given by the simplified relation for a disk:

$$I = \frac{1}{2} m \left(r_{ext}^2 - r_{int}^2 \right) \tag{8.17}$$

where m is the mass, r_{ext} is the external radius and r_{int} is the internal radius of the cylinder.

Normally, flywheels are designed for uninterruptible power supply (UPS) systems where power is needed for a short period of time. An example of a small-scale energy storage solution is a flywheel equipped with a maximum power capacity of 300 kW and a maximum energy storage of 4 MW s at 100 kW, in a 2 m-high container with a footprint of 0.6 m^2. For higher storage requirements, larger dimensions and more containers are needed.

A main advantage of flywheel storage is its capacity to sustain practically unlimited charging and discharging cycles without degradation. Moreover, it is highly reliable and more durable compared to VRLA electrochemical battery storage, with a service life expectancy of at least 20 years. The use of inert materials also renders flywheels more environmental-friendly than traditional batteries. However, there are several disadvantages as well that can significantly influence certain applications. First of all, flywheels still cannot be immediately used in non-stationary applications because of their heavy weight and the gyroscopic effect. Such effect can completely change the dynamical behaviour of the vehicles and cause issues in mechanical design. Another limitation is related to the safety risk when a flywheel may explode due to excess mechanical stress. It is therefore vital to consider the resistance capacity in the flywheel design in order to avoid any involuntary overcharging.

Instead of mechanical bearings, magnetic bearings are adopted in flywheel storage in order to limit the friction forces that reduce the storage capacity. Superconductivity is also employed to improve the non-contact bearing with the use of levitation force, so that the system can obtain a very high rotation regime as well as a high power-to-volume ratio.

8.9 Superconducting Magnetic Energy Storage

Apart from flywheels, superconductivity is also applied in the *Superconducting Magnetic Energy Storage* (SMES). In this technology, a superconducting magnet is kept at cryogenic temperatures and energy is stored by electromagnetic induction, expressed by:

$$E = \frac{1}{2} LI^2 \tag{8.18}$$

and power by:

$$P = \frac{dE}{dt} = UI \tag{8.19}$$

where L, I, U are the coil inductance, direct current and direct voltage. Since the energy is stored in the form of circulating current, the system response time can be very short; however the need to convert the direct into alternate current can reduce the system efficiency to the range of 95% with a power density close to 10^3 W/kg.

References

1. Besenhard J O, Eichinger G (1976) High Energy Density Lithium Cells. Part I. Electrolytes and Anodes. J Electroanal Chem 68 (1):1–18
2. Besenhard J O, Eichinger G (1976) High Energy Density Lithium Cells. Part II. Cathodes and Complete Cells. J Electroanal Chem 72(1):1–31
3. Bilodeau A, Agbossou K (2006) Control analysis of renewable energy system with hydrogen storage for residential applications. Journal of Power Sources 162:757–764
4. Conway B E (1999) Electrochemical Supercapacitors, Plenum Publishing, New York
5. Huang K-L, Li X, Liu S, Tan N, Chen L (2008) Research progress of vanadium redox flow battery for energy storage in China. Renewable Energy 33:186–192
6. Jin J X (2007) HTS energy storage techniques for use in distributed generation systems. Physica C 460–462:1449–1450
7. Koshizuka N, Ishikawa F, Nasu H, Murakami M et al (2003) Progress of superconducting bearing technologies for flywheel energy storage systems. Physica C (386):444–450
8. Kötz R, Carlen M (2000) Principles and applications of electrochemical capacitors. Electrochimica Acta 45:2483–2498
9. Lee K, Yi J, Kim B, Ko J et al (2008) Micro-energy storage system using permanent magnet and high-temperature superconductor. Sensors and Actuators A 143:106–112
10. Linden D, Reddy T B (eds.) (2002) Handbook Of Batteries, 3rd ed. McGraw-Hill, New York
11. Martha S K, Hariprakash B, Gaffoor S A, Amblavanan S et al. (2005) Assembly and performance of hybrid-VRLA cells and batteries. Journal of Power Sources (144) 2:560–567

12. Muneret X, Gobé V, Lemoine C (2005) Influence of float and charge voltage adjustment on the service life of AGM VRLA batteries depending on the conditions of use. Journal of Power Sources 144 (2):322–328

13. Pascoe P E, Anbuki A H (2004) A VRLA battery simulation model. Energy Conversion and Management 45 (7–8):1015-1041

14. Peltier R (2011) Energy storage enables just-in-time generation. Power, April

15. Sharma P, Bhatti T S (2010) A review on electrochemical double-layer capacitors. Energy Conversion and Management 51:2901–2912

16. Silberberg M (2006) Chemistry: The Molecular Nature of Matter and Change, 4th ed. McGraw-Hill Education, New York

17. Skyllas-Kazacos M, Rychcik M, Robins R et al (1986) New all-vanadium redox cell. J Electrochem Soc 133:1057–1058

18. Sum E, Skyllas-Kazacos M (1985) A study of the V(II)/V(III) redox couple for redox flow cell applications. J Power Sources 15:179–190

19. Sum E, Rychcik M, Skyllas-Kazacos M (1985) Investigation of the V(V)/V(IV) system for use in the positive half-cell of a redox battery. J Power Sources 16:85–95

20. Sung T H, Han S C, Han Y H, Lee J S et al (2002) Designs and analyses of flywheel energy storage systems using high-Tc superconductor bearings. Cryogenics 42:357–362

21. Ulleberg Ø (1998) Stand Alone Power Systems for the Future: Optimal Design, Operation and Control of Solar Hydrogen Energy Systems. Ph.D. thesis, Norwegian University of Science and Technology, Trondheim

22. Whittingham M S (1976) Electrical Energy Storage and Intercalation Chemistry. Science 192 (4244): 1126–1127

23. Wolsky A M (2002) The status and prospects for flywheels and SMES that incorporate HTS. Physica C (372–376):1495–1499

24. Zheng J P, Jow T R (1996) High energy and high power density electrochemical capacitors. Journal of Power Sources 62:155–159

9

Study and Simulation of Solar Hydrogen Energy Systems

Solar hydrogen energy systems integrate technologies of solar and wind energies to deliver a reliable energy supply. To achieve this, the volatile energy input from the sun and the wind must be smoothed out by the use of energy storage devices. The conversion of renewable energy to electric energy is performed by photovoltaic installations or aerogenerators, while the change of electricity to hydrogen and the reverse procedure are carried out by electrolysers, storage apparatus, fuel cells along with other satellite devices that ensure the efficient functioning of the systems. Mathematical modelling and the related simulations are hereby detailed to better understand and design the complete energy system.

9.1 Solar Hydrogen Energy Systems

A solar hydrogen energy system is a set of several sub-systems with different technologies combined to work together as a whole. It converts energy from a solar renewable source to store it as hydrogen, and transforms it into electric energy to supply a load. The complete system schematic is depicted in Figure 9.1.

In a general configuration, the renewable energy source is connected by a boost-converter[1] to the DC bus-bar that works as the backbone of all the system. Through a buck-converter[2], the bus-bar supplies power to the electrolyser to produce hydrogen and oxygen that are stored later. When the storage needs to be controlled with thermal cycles, the control system operates with other satellite sub-systems that are characterised by a *Coefficient of Performance* (COP) when they heat up or cool down the storage vessels. The COP represents the ratio between the thermal energy provided by the storage sub-systems and the electric energy employed to provide the thermal cycle. The higher the COP is, the higher the overall system efficiency becomes.

---

[1] A *boost-converter* is a DC-DC power supply device that provides an output voltage higher than its input voltage.

[2] A *buck-converter* is a DC-DC power supply device that provides an output voltage lower than its input voltage.

Zini G., Tartarini P.: Solar Hydrogen Energy Systems. Science and Technology for the Hydrogen Economy. DOI 10.1007/978-88-470-1998-0_9, © Springer-Verlag Italia 2012

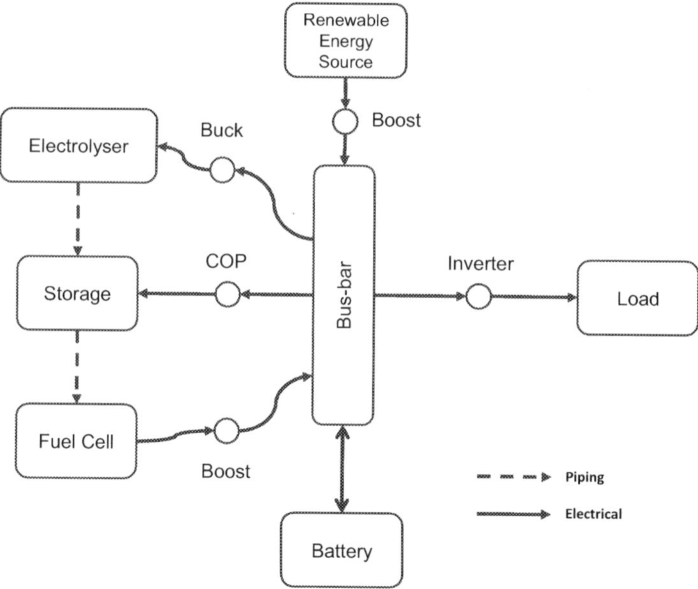

Fig. 9.1. Schematic of a solar hydrogen energy system

A battery is connected to the bus-bar to smooth out the voltage glitches that can be caused by the transients when the sub-systems are switched on and off. The battery also assists as an initial power supply before the fuel cell is turned on. Depending on the chosen control logic, the battery can be discharged only until it reaches its lowest SOC or even before. Finally, the load is supplied by the bus-bar through a DC/AC converter (inverter).

The storage method can be based either on traditional or on advanced technologies. In this chapter, the simulation of the solar hydrogen energy system will adopt traditional compression storage and physisorption on activated carbons.

9.2 Control Logic

Since all the devices in the system introduce losses according to their respective operating characteristics, the overall efficiency loss of the whole system can become significant. To avoid excessive losses and to achieve the highest possible overall efficiency, the system control logic must be optimised.

To coordinate such a complex energy system, control devices like a PLC or an industrial-grade PC are required to monitor the operating conditions of all the sub-systems and to make decisions regarding the overall system management. A flow-chart of the decision-making process can be devised as in Figure 9.2.

The process is based on the balancing of the bus-bar currents:

$$I_{res \to bus} - I_{bus \to el} - I_{bus \to sto} + I_{fc \to bus} - I_{bus \to load} \pm I_{bus \leftrightarrow bat} \qquad (9.1)$$

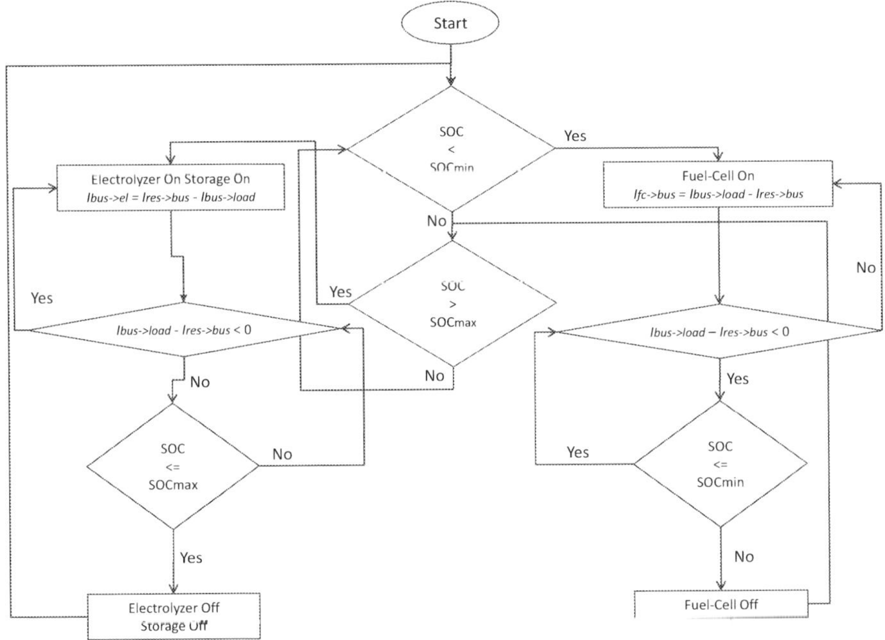

Fig. 9.2. Control logic flow chart (Reproduced with permission from [29])

where the input currents are positive, the output currents are negative and the subscripts are as follows:

- *res*: renewable energy source;
- *bus*: bus-bar;
- *el*: electrolyser;
- *sto*: storage control system;
- *fc*: fuel cell;
- *load*: load;
- *bat*: battery.

When the renewable source does not convert enough energy to supply the load (when $I_{res \to bus}$ is lower than $I_{bus \to load}$), the battery discharges a positive $I_{bat \leftrightarrow bus}$ to fill in the power gap. When the battery reaches a pre-determined lower SOC, the fuel cell is turned on to assist with power supply. When $I_{res \to bus}$ is higher than $I_{bus \to load}$, the fuel cell is disconnected and the batteries are recharged with a negative $I_{bat \leftrightarrow bus}$. When the batteries reach the upper SOC, they are disconnected and the electrolyser and the storage control sub-system are activated to store hydrogen.

Alternative control logics can be devised with various strategies to optimize system operations.

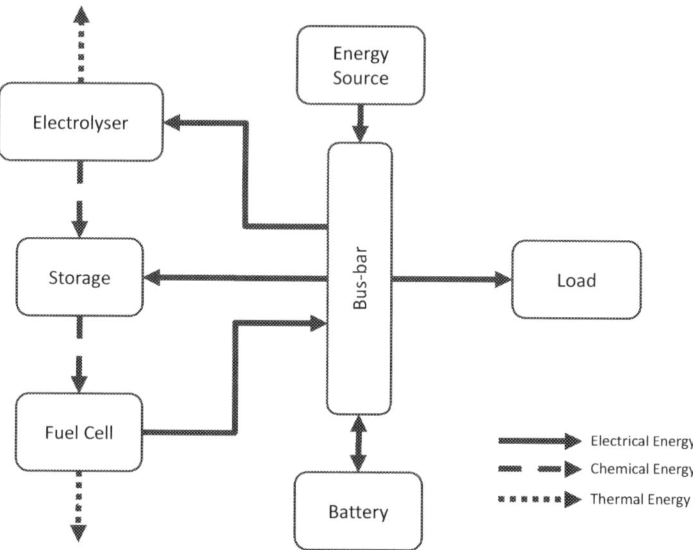

Fig. 9.3. Energy flows in the hydrogen system

9.3 Performance Analysis

The renewable, chemical, thermal and electric energy flows are shown in Figure 9.3 to help evaluate the efficiencies of the sub-systems as well as the overall system performance. This is important to improve the technical design, to deliver a correct cost/benefit analysis and to evaluate the system viability for real-life uses.

When the performance is evaluated, the thermal energy coming in and out of the activated carbon storage system is not considered, since what is provided to the storage system during desorption is obtained during adsorption as a first approximation. In traditional compression storage, thermal flows are lost to the environment.

No heat from the fuel cell or the electrolyser operations is considered in the calculations. If a fuel cell is used as a CHP unit, the final overall system efficiency will be higher.

9.3.1 Sub-Systems Efficiencies

9.3.1.1 Photovoltaic Modules

The conversion efficiency of a photovoltaic field depends on the technology of the modules and on the various losses that reduce the original photovoltaic cell efficiency. The efficiency in the following simulations will be considered at 12.7%, a value in the lower range of crystalline silicon PV technology.

The efficiency reduction from this part is the first and the most drastic among all the factors contributing to the overall system efficiency loss, therefore the improvement in the photovoltaic efficiency will also dramatically improve the final system performance. Although the conversion efficiency of crystalline silicon cells seems to have reached its maximum, around 20–25 %, other PV technologies are projected to deliver higher conversion ratios potentially as high as 40–50% in the years to come.

9.3.1.2 Aerogenerator

The kinetic energy in the wind flow that hits the aerogenerator rotor is converted into electric energy with power losses caused by air flow turbulences, mechanical couplings, asynchronous generator operation and DC/AC conversion. These losses must be considered in order to obtain the final efficiency as well as the cut-in, the cut-out and the continuous speeds.

The values used in the simulation runs are given in Table 9.1.

Table 9.1. Wind system conversion efficiencies

c_p	0.45
Mechanical Losses	60%
R1, R2	0.02
DC/AC Conversion Efficiency	90%

The efficiency η_{Wind} is the ratio between the power P_{Aero} converted in electric energy and the power in the wind flow intercepted by the rotor:

$$\eta_{Wind} = \frac{P_{Aero}}{\frac{\rho}{2} A \int_0^T v^3 dt}. \tag{9.2}$$

9.3.1.3 Electrolyser

The electrolyser efficiency is defined as:

$$\eta_{EL} = \frac{LHV_{H_2} \int_0^T \dot{n}_{H_2} dt}{\int_0^T P_{el} dt} \tag{9.3}$$

where LHV_{H_2} is the hydrogen lower heating value, $\dot{n}_{H_2}$ is the hydrogen molar flow and P_{el} is the input power to the electrolyser. The integrals are computed between 0 and T in seconds/year.

The Faraday's efficiency η_F is defined by:

$$\eta_F = \frac{\dot{n}_{H_2} z F}{N_c I_{el}} \tag{9.4}$$

where $\dot{n}_{H_2}$ is the hydrogen molar flow, N_c is the number of cells in series, I_{el} is the current, z is the number of electrons in the water-splitting reaction (equal to 2) and F is the Faraday's constant. In the simulations performed in the next sections, F is assumed to be 90%.

9.3.1.4 Fuel Cell

The fuel cell efficiency is defined by:

$$\eta_{FC} = \frac{U_{fc}}{N_c U_{rev}} \tag{9.5}$$

where U_{fc} is the voltage at the fuel cell electrodes, N_c is the stack cell number and U_{rev} is the reversible potential of a stack cell.

The Faraday's efficiency η_F is defined by:

$$\eta_F = \frac{N_c I_{fc}}{\dot{n}_{H_2} z F} \tag{9.6}$$

where I_{fc} is the current yielded by the fuel cell. Also in this case, F is assumed to be 90%.

9.3.1.5 Compressor

The compression efficiency is defined as:

$$\eta_{comp} = \frac{\dot{n}_{gas} L_{comp,id}}{P_{comp}} \tag{9.7}$$

where $\dot{n}_{gas}$ is the molar flow of the gas undergoing the compression process, $L_{comp,id}$ is the ideal molar compression work and P_{comp} is the power requested by the compressor. In this case, the compression efficiency is approximated as constant and equal to 92%.

9.3.1.6 Electric Systems

The electric satellite systems that assist the main operation of the hydrogen system are the buck-converter, the boost-converter and the the DC/AC converter. Their efficiencies depend on the respective operating currents and voltages and are given by the manufacturers' datasheets. The values given in Table 9.2 have been used in the simulation runs.

Table 9.2. Electric satellite systems efficiencies

Buck-converter	95%
Boost-converter	95%
Inverter	95.5%

9.3.2 Complete System Efficiencies

The evaluation of the complete system efficiencies is fundamental as it indicates how much renewable energy source has been converted in chemical energy and in electric energy supplied to the load. These calculations are also important in the design phase since they give the possibility to optimise the control logic and system dimensioning.

In the following, the efficiency calculations do not consider the thermal energy flows to and from the electrolyser and the fuel cell and do not include either the electric energy needed by the satellite systems (such as the programmable control logic or the system actuators).

9.3.2.1 Hydrogen Production Efficiency

If the renewable energy source is solar photovoltaic energy, the production efficiency is computed as:

$$\eta_{Prod} = \frac{HHV_{H_2} \int_0^T \dot{n}_{H_2} dt}{A \int_0^T G_T dt} \tag{9.8}$$

where HHV_{H_2} is the hydrogen higher heating value, A is the photovoltaic field total area and G_T is the global solar irradiance on the surface of the photovoltaic modules.

If the renewable energy source is wind energy, the production efficiency is computed as:

$$\eta_{Prod} = \frac{HHV_{H_2} \int_0^T \dot{n}_{H_2} dt}{\frac{\rho}{2} A \int_0^T v^3 dt} \tag{9.9}$$

where ρ is the air density and v is the wind speed on the rotor that sweeps across a surface A.

9.3.2.2 Direct Route Efficiency

The *direct route* efficiency η_{DR} is given by:

$$\eta_{DR} = \eta_{Boost,PV} \, \eta_{Inv} \tag{9.10}$$

where $\eta_{Boost,PV}$ is the boost converter efficiency and η_{Inv} is the inverter efficiency. It represents the conversion efficiency of the renewable energy source when it directly supplies the load.

9.3.2.3 Hydrogen Loop Efficiency

The *hydrogen loop* efficiency η_{HL} is defined by:

$$\eta_{HL} = \eta_{Buck,El} \, \eta_{El} \, \eta_{FC} \, \eta_{Boost,FC} \, \eta_{Inv} \tag{9.11}$$

where $\eta_{Buck,El}$ is the efficiency of the converter supplying the electrolyser, η_{El} is the electrolyser efficiency, η_{FC} is the fuel cell efficiency, $\eta_{Boost,FC}$ is the converter efficiency out of the fuel cell and η_{Inv} is the inverter efficiency. It is the conversion efficiency of the renewable energy sources along the path electrolyser $\rightarrow$ fuel cell $\rightarrow$ load.

9.3.2.4 Complete System Efficiency

The complete system efficiency η_{Sys} when the solar photovoltaic energy is applied is given by:

$$\eta_{Sys} = \frac{\eta_{DR} \int_0^T \left(P_{PV} - P_{Buck,El} - P_{St} \right) dt + \eta_{HL} \int_0^T P_{Buck,El} dt}{A \int_0^T G_T dt}. \tag{9.12}$$

And in the case of wind energy the calculation becomes:

$$\eta_{Sys} = \frac{\eta_{DR} \int_0^T \left(P_{Aero} - P_{Buck,El} - P_{St} \right) dt + \eta_{HL} \int_0^T P_{Buck,El} dt}{\frac{\rho}{2} A \int_0^T v^3 dt} \tag{9.13}$$

where P_{PV} is the power from the photovoltaic field, P_{FC} is the power obtained from the fuel cell and P_{St} is the power absorbed by the storage system. η_{DR}, η_{HL}, P_{Aero}, $P_{Buck,El}$ have already been previously defined.

These values represent the conversion efficiencies of the whole system, from the renewable energy source to its final use to the load. They are the most important indicators to help evaluate the system performances and modify the design. The higher these values are, the better the entire system performs.

9.4 Simulation with PV Conversion and Compression Storage

The overall system behaviour and performance can be simulated and studied by integrating all the mathematical models described in the previous chapters for the different sub-systems. The system architecture is the same as in Figure 9.1. In this case, the renewable energy chosen is photovoltaic energy with solar radiation as the source for the overall system. Its modelling has been introduced in Chapter 4 and the behaviour of the electrolyser and the fuel cell has been detailed in Chapter 3. The compressed hydrogen storage is presented in Chapter 7, while Chapter 8 examines the electrochemical battery that smooths the sub-systems on/off cycling and maintains a constant voltage on the bus-bar.

The load profile chosen to simulate the system behaviour is a typical profile of domestic single or multiple housings in southern California, U.S.A. The annual electric energy consumption is 6466.7 kWh. The same location receives an average daily horizontal global radiation of 18.05 MJ/m^2 and its annual variation (with a time scale of 1 h) is also input in the simulation [29].

To obtain the current delivered to the load, it is necessary to introduce a loss caused by the inverter efficiency η_{inv}. The current is given by:

$$I_{bus \to load} = \frac{P_{bus}}{\eta_{inv} U_{bus}} \tag{9.14}$$

where P_{bus} and U_{bus} are the power and voltage of the bus-bar.

The system data are reported in Table 9.3.

Table 9.3. System data

System	Type	Parameter	Value	Parameter	Value
PV	Polycristalline	Peak power	167 kW$_p$	$I_{0,ref}$	1 A
		No. modules	32	$I_{l,ref}$	8.1 A
		Efficiency	12.7%	I_{sc}	8.02 A
		U_{oc}	29.04 V	U_{mp}	22.97 V
		N_s	48	I_{mp}	7.27 A
		μ_{Isc} (A/°C)	34×10^{-5}	$R_{sh,ref}$	700 Ω
		$R_{sh,ref}$	0.2 Ω	Azimuth	South
		Tilt	30 °		
EL	SPE	$\eta_{F,el}$	90%	$U_{el,0}$	22.25 V
		$C_{1,el}$ (°C^{-1})	−0.1765	$I_{el,0}$	0.1341 A
		$C_{2,el}$ (V)	5.5015	R_{el}	-3.3189 Ω
		N_{cells}	24		
FC	PEM	$\eta_{F,fc}$	90%	N_{fc}	35
		$C_{1,fc}$ (°C^{-1})	−0.013	$U_{fc,0}$	33.18 V
		$C_{2,fc}$ (V)	−1.57	$I_{fc,0}$	8.798 A
		R_{fc}	-2.04 Ω°C		
Battery	VRLA	E_{max}	36 MJ	V_{B0}	48 V
		E_{in}	18 MJ	R_i	0.076 Ω
		α	0	SOC_{min}	25%
		K_i	0	SOC_{max}	85%
Compression Storage		Polytropic	1.475	Efficiency	92%
		H$_2$ initial storage (mol)	3000	Storage volume (m^3)	0.8
Electrics		$\eta_{BoostConv}$	95%	$\eta_{BuckConv}$	95%
		$\eta_{Inverter}$	95.5%		

The simulation results show that the solar hydrogen energy system is capable of working independently from the grid in stand-alone mode. The global solar radiation on the inclined surface (30°, azimuth south) is given in Figure 9.4. The photovoltaic system converts this radiation into to electric power as demonstrated in Figure 9.5, which also clearly illustrates seasonality.

The annual hydrogen production is shown in Figure 9.6: the spring productivity is over 7×10^{-3} mol/s while the productivity in winter is higher than 5×10^{-3} mol/s. The electric power supplied to the electrolyser to obtain hydrogen is depicted in Figure 9.7.

The annual hydrogen consumption in the fuel cell (Figure 9.8) and the electric energy yield of the cell (Figure 9.9) are highly dependent on the meteorological and the load conditions. For example, between the 220th and 240th day of the year considered for the simulation, the weather conditions have been characterised by a solar radiation lower than usual, hence the hydrogen consumption is augmented to compensate for the diminished photovoltaic productivity.

The production of hydrogen clearly demonstrates a seasonality with increases in spring and decreases in winter (Figure 9.10).

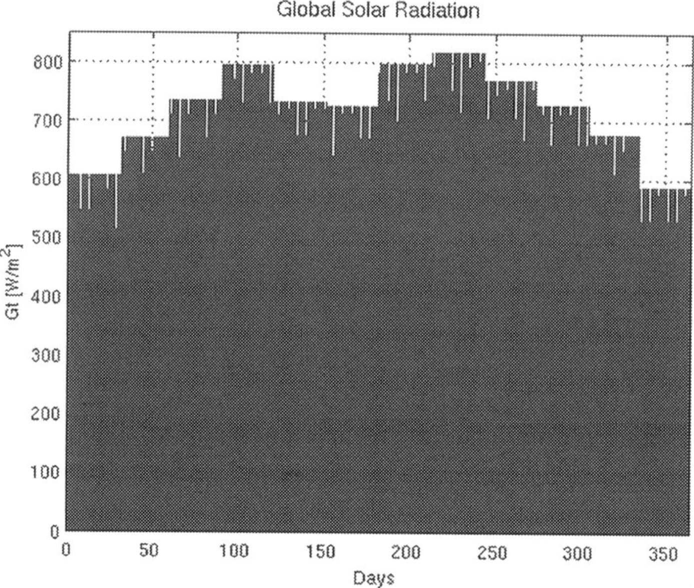

Fig. 9.4. Global solar radiation in one year

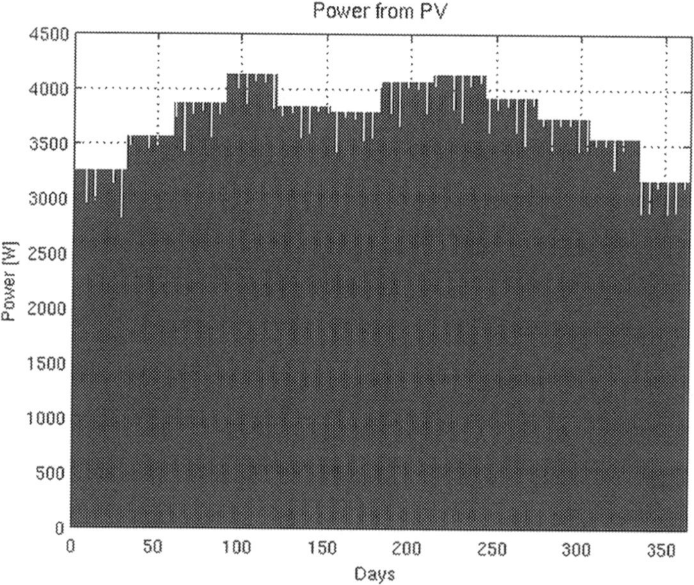

Fig. 9.5. Electric power converted by the PV system

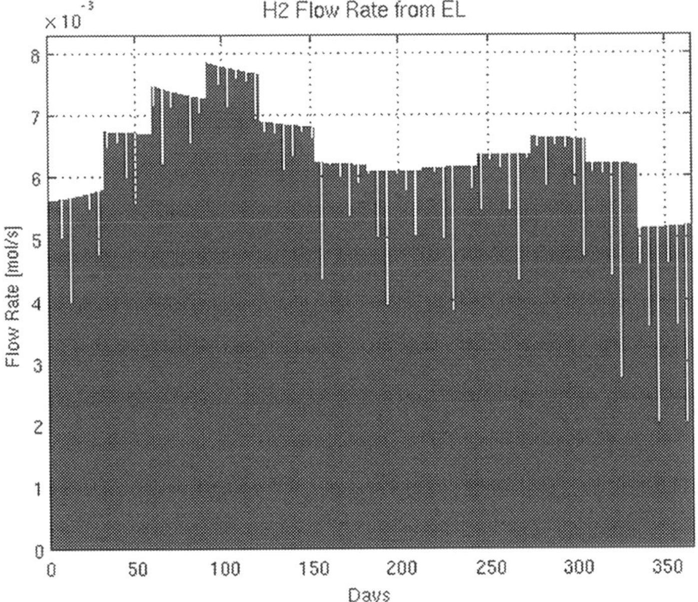

Fig. 9.6. Annual hydrogen production

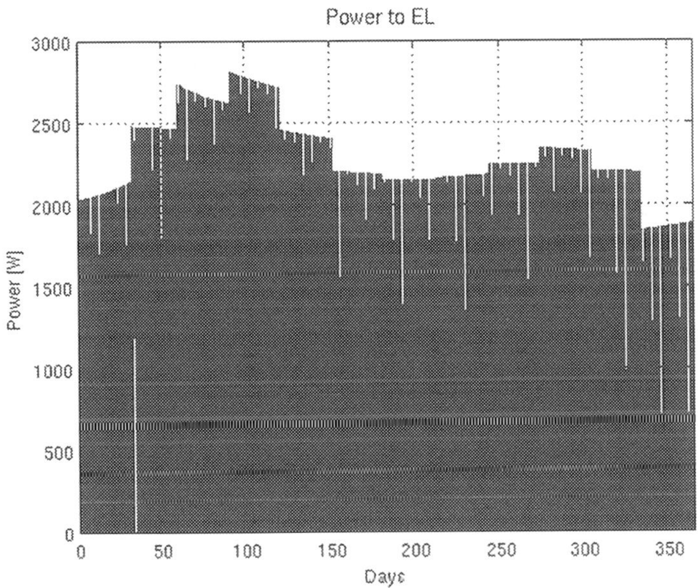

Fig. 9.7. Electric power supplied to the electrolyser in one year

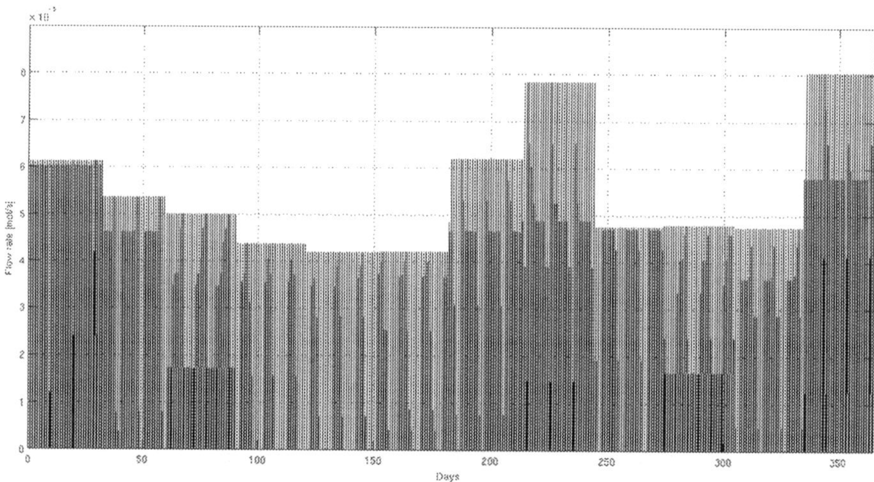

Fig. 9.8. Hydrogen flow to the fuel cell in one year (Reproduced with permission from [29])

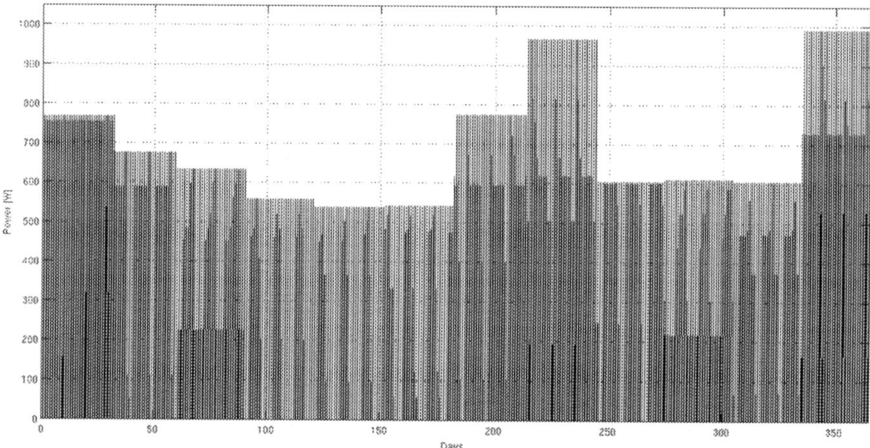

Fig. 9.9. Electric power yielded form the fuel cell in one year (Reproduced with permission from [29])

At the end of the year, the storage has a surplus of 6.62 kg with respect to the initial charge. This demonstrates that the system is capable of working in the stand-alone mode, granting the engineers the possibility to explore other potential uses of this excess hydrogen, such as refuelling hydrogen-based vehicles. The peak hydrogen quantity of 19.69 kg is reached on day 183, while the minimum of 1.21 kg falls on day 32.

The pressure variation trend is illustrated in Figure 9.11 with the maximum pressure being close to 30 MPa (300 bar). The oxygen mass and pressure trends are similar to those of hydrogen.

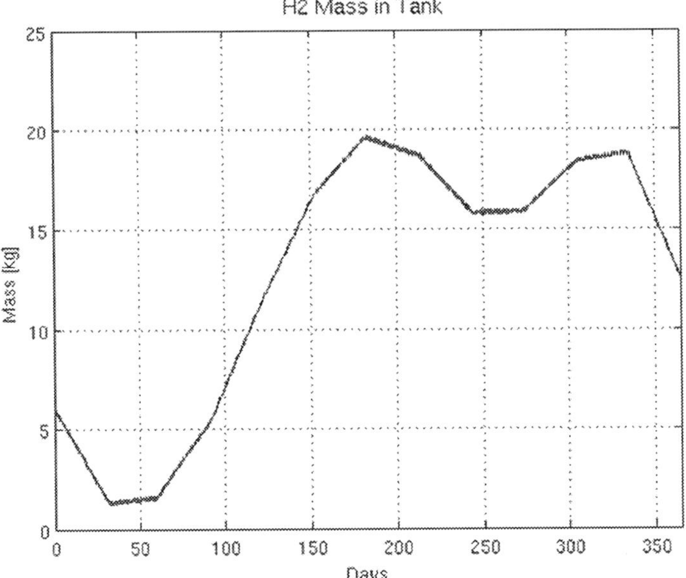

Fig. 9.10. Hydrogen mass variation in the tank in one year

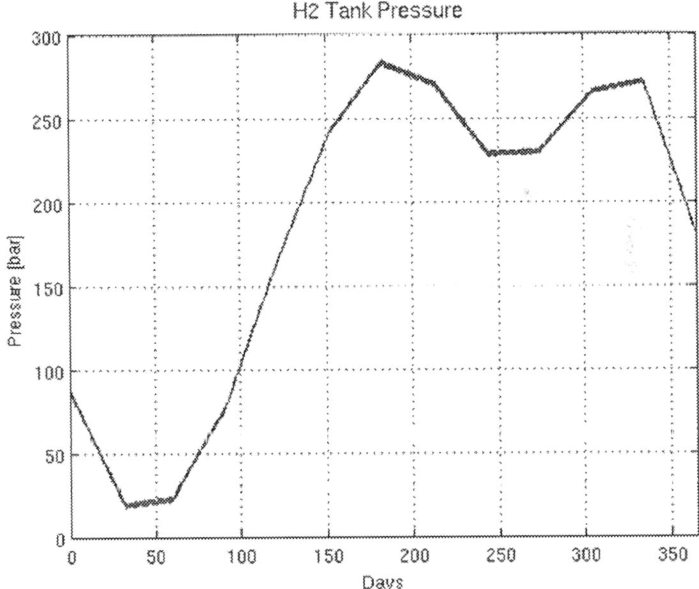

Fig. 9.11. Hydrogen pressure variation in the tank in one year

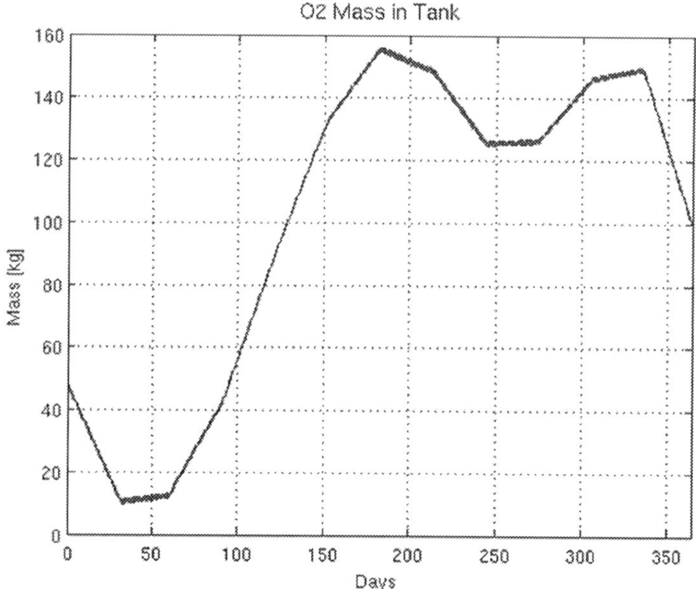

Fig. 9.12. Oxygen pressure variation in the tank in one year

Figure 9.12 demonstrates the power supplied to the load. The maximum power is requested during the summer due to high amount of energy consumed for cooling. The load is completely supported by the solar hydrogen energy system with no need of power extracted from the grid.

The efficiencies and other performance data are reported in Table 9.4.

Table 9.4. Yearly performance with compression storage, PV conversion

Solar Radiation	6.75 GJ/m^2
kWh/kWp	1921
kg H$_2$/kW$_p$	14.84
η_{FC}	63.1%
η_{EL}	66.8%
η_{Prod}	3.8%
η_{Sys}	8.5%
Minimum storage (day 32)	1.21 kg (602 mol)
Maximum storage (day 183)	19.69 kg (8.27 × 10^3 mol)
End of period H$_2$ excess capacity	6.62 kg (3.28 × 10^3 mol)
η_{Prod}	4.0% (H$_2$ compression only)
η_{Sys}	9.3% (H$_2$ compression only)
Minimum storage (day 32)	1.34 kg (6680 mol) (H$_2$ compression only)
Maximum storage (day 335)	21.89 kg (10.86 × 10^3 mol) (H$_2$ compression only)
End of period H$_2$ excess capacity	9.76 kg (4.84 × 10^3 mol) (H$_2$ compression only)

The complete system efficiency η_{Sys} is 8.5%. As already discussed, the greatest energy loss is attributed to the low photovoltaic conversion rates. Advances in photovoltaic technology will reduce such losses and largely improve the prospects of the solar hydrogen energy system.

9.5 Simulation with PV Conversion and Activated-Carbon Storage

The same system simulated in Section 9.4 is examined here by employing activated-carbon storage instead of traditional compressed hydrogen [45]. The mathematics of activated carbon storage have been introduced in Chapter 7. All the other subsystems, the meteorological and the load specifications remain the same as in Section 9.4.

In this case, hydrogen is stored by physisorption in vessels containing activated carbons where the gas is absorbed as mono-layers over the solid carbon adsorbent surface. This kind of adsorption phenomenon can be modelled as a type 1 isotherm at supercritical temperatures.

Instead of using the classical Langmuir's equation which assumes homogeneous adsorbent surfaces, the *Langmuir-Freundlich's* (L-F) equation can be employed here, since it is better suited for solids with non-homogeneous surfaces like the activated carbons. The L-F equation defines the *absolute adsorption* n^S [mol] as:

$$n^S = n^0 \left(\frac{bp^q}{1 + bp^q} \right) \tag{9.15}$$

where n^0 is the *saturated adsorption* capacity, p is the pressure applied to the gas-solid system, b is a parameter accounting for energy change in adsorption (which climbs as temperature decreases) and q is an exponent parameter ranging between 0 and 1 and associated with the surface heterogeneity (which increases with temperature at supercritical conditions).

The n^0, b and q coefficients obtained with non-linear regressions on absolute adsorption isotherms are reported for the powdered AX-21 activated carbons in [38]→[42]. Such carbon structures have a very high specific surface of 2745 m^2/g and a density of 0.3 kg/l. Such isotherms can be obtained by measuring adsorption and desorption over a temperature range of 77–298 K (at 20 K intervals) and a pressure range of 0 to 7 MPa. A standard volumetric method and a special cryostat are used for adsorbent samples of about 20 g. The correlation coefficients between the referenced data and the obtained L-F empirical model estimates are higher than 0.999.

Absolute adsorption alone is not the only source of contribution to the total moles of hydrogen stored in the activated carbon. From the Gibbs's theory of adsorption, if V_a is the volume of hydrogen in the adsorbed phase, there remains a volume of V_g available for the non absorbed hydrogen that stays in its gas phase. The total volume available for hydrogen storage V_{tv} is actually:

$$V_{tv} = V_a + V_g \tag{9.16}$$

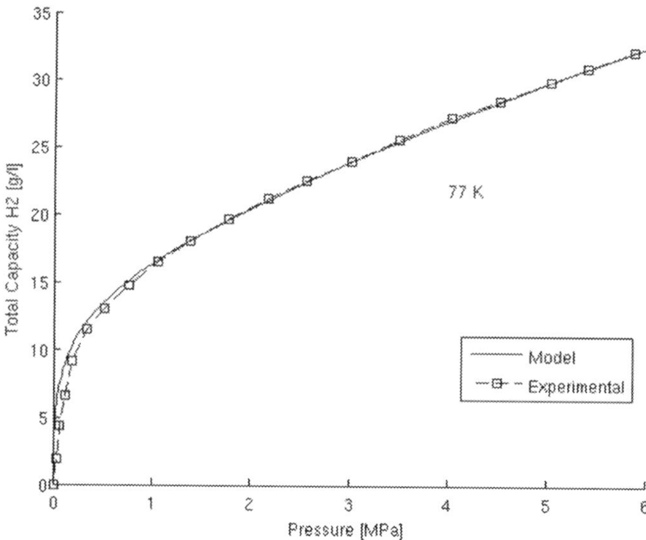

Fig. 9.13. Total capacities and comparison with experimental data (Reproduced with permission from [45])

and the total capacity of hydrogen storage is given both by physisorption and by compression:

$$C_{tot} = n^S + \rho_g V_g \tag{9.17}$$

where ρ_g is the hydrogen gas density.

To compute C_{tot}, the moles of gas stored by compression only in the free volume where adsorption does not occur must be added to n^S, already obtained by non linear regression. From the experimental data at 6 MPa, 77 K, C_{tot} is 32.5 g/l and n_S is 17.7 g/l; their difference represents the quantity of gas stored by compression only in one unit volume, equal to 14.8 g/l (7.34 mol/l). Solving for V in the Van der Waals' law for hydrogen at 6 MPa, 77 K and 7.34 mol/l yields $V_g = 0.763385$ l, that will be used to compute the moles stored by compression at every p-T pair.

Figure 9.13 is the comparison of the total capacity between the data reported in related literature and the results provided by the model. The fit has a correlation coefficient of 0.995. The calculation from the literature data demonstrates that at 6 MPa the gravimetric storage is 10.8% and the volumetric storage is 32.5 g/l.

Figure 9.14 depicts the total capacity isotherms for five temperature values.

Since adsorption and desorption kinetics are fast with no signs of hysteresis and adsorption enthalpies are relatively low, fast hydrogen charging and discharging can happen along the isotherm curves. A possible operating cycle can be devised to be composed of four transformations: *isobar pre-charging* at 0.1 MPa when AC are cooled from 153 K to 77 K and hydrogen starts to be adsorbed; *isothermal charging* at 77 K when the pressure is raised from 0.1 MPa to 6 MPa and the adsorption is completed; *isobar pre-discharging* at 6 MPa when the temperature increases from

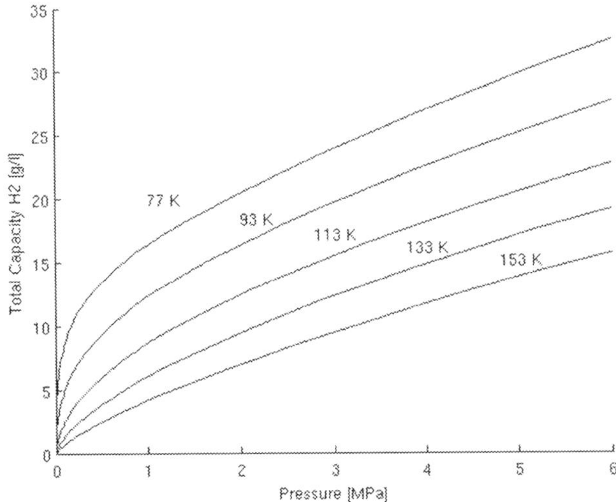

Fig. 9.14. Total capacity isotherms in the range 77–153 K (Reproduced with permission from [45])

77 K to 153 K and hydrogen desorption starts, and finally *isothermal discharging* at 153 K from 6 MPa to 0.1 MPa when the desorption is completed (Figure 9.15).

Due to the nature of the working cycle, the tanks must be repeatedly heated and cooled. The storage components can be designed as a cluster of several different vessels or tanks that can also function individually. Each tank, filled with AX-21

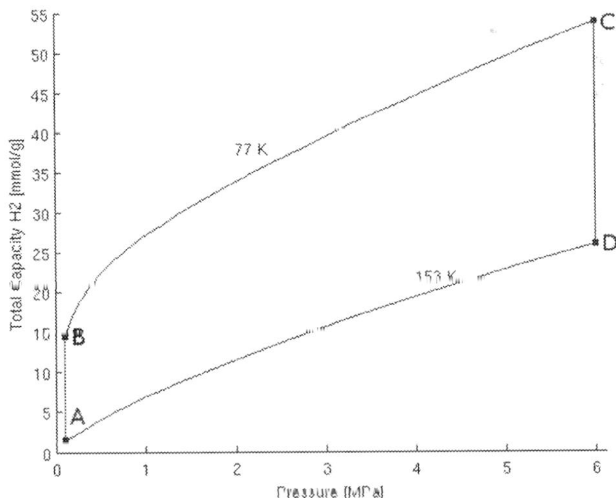

Fig. 9.15. Adsorption/desorption working cycle for AC storage (Reproduced with permission from [45])

powder, can be devised as a Dewar-like cylinder with elliptical end-caps. An outer vessel equipped with a vacuum chamber and multi-layered radiation shields encases the inner cylinder in order to minimize the overall wall heat transmittance. Inside the inner cylinder, a conduit drives the fluids suitable for cooling or heating the AC to the desired temperatures. The final vessel design is outside the scope of the simulation, but its physical characteristics must be known in order to compute the heat flows and the relevant energy employed by the thermal cycles running during the storage operations.

Adsorption is a spontaneous exothermic reaction. The isosteric adsorption enthalpy is estimated from the *Clausis-Clapeyron's* equation obtained through the experimental data of hydrogen adsorption. The resulting equation yields the heat which is released during adsorption or required during desorption:

$$-\Delta H_{ads} = RT^2 \left[\frac{\ln p}{b} (0.009684 \times \ln p - 0.10595) \right] \tag{9.18}$$

with b given by regression on experimental data:

$$b = 7.848 - 0.10595T. \tag{9.19}$$

To estimate the thermodynamic behaviour of the tanks during the working cycle, four different phases are considered: pre-charging (A-B transformation), charging (B-C), pre-discharging (C-D) and discharging (D-A). Charging and discharging are two procedures that do not occur at the same time: a tank is charged only if it has reached point A and is discharged only after it has reached point C. In the isobaric A-B transformation, the thermal balance equation is given by:

$$\dot{Q}_{ads} + \dot{m}_{H_2} c_{p,H_2} \left(T_f - T_i \right) + \left(m_{H_2,ini} c_{p,H_2} + m_{AC} c_{p,AC} \right) \frac{dT}{dt} + \dot{Q}_{loss} = \dot{Q} \tag{9.20}$$

where $\dot{Q}_{ads}$ is the derivative with respect to time of the isosteric heat of adsorption or desorption, $\dot{m}$ is the mass flow rate of hydrogen (subscript H_2), T_i is the initial temperature of hydrogen ranging between 289-300 K and computed during the simulations run by the EL sub-system, T_f is the final hydrogen temperature, $m_{H_2,ini}$ is the number of hydrogen moles in the tank at the p-T values at the start of charging and c_p is the specific heat for hydrogen (subscript H_2) and activated carbons (subscript AC). $\dot{Q}_{loss}$ is the thermal power lost to the environment, $\dot{Q}$ is the thermal power that has to be managed by the system controlling the complete charging-discharging cycle.

To simplify the simulation without sacrificing precision, the heat control devices are represented by its coefficient of performance (COP), hereby assumed equal to 1. The number of storage tanks is assumed to be five in this simulation.

The isobaric pre-charging is performed with a gradual and slow variation in temperature and it takes approximately one day to complete. The B-C transformation has the same equation except for the contribution of activated carbon bulk storage, which is zero since the B-C transformation is isothermal. The pre-discharging isobaric C-D timing is determined by the downstream fuel cell requirements. In the pre- and discharging transformations C-D-A, the equations are the same, apart from the

Table 9.5. AC storage system data (Reproduced with permission from [45])

Activated carbons type	AX-21
Specific surface	2745 m^2/g
Density	0.3 kg/l
Number of tanks	5
Tank type	dewar, multi-layer radiation shields
Single tank storage	3870 mol
Total initial charge	7970 mol
External radius	250 mm
Length	1.223 mm
Inner vessel wall width	4 mm
Outer vessel wall width	2 mm
Vacuum cavity width	36 mm
Vacuum thermal conductivity	0.004 W/mK
Radiation shields (multi-layered)	20
Operating Temperatures	77–153 K
Cooling Fluid	N

hydrogen moles at the start of discharging, the C-D derivative of T over time and the final temperature for hydrogen. To operate at such a low temperature the fluid of choice is nitrogen. Physisorption occurs at 77 K while the temperature can be as low as 22 K for liquefaction storage. Since no hydrogen boil-off is possible at the physisorption temperatures as hydrogen is in its gaseous state, the use of nitrogen as the cryogenic fluid renders the operation easier and cheaper than liquefaction.

The hydrogen pressure cycles of the transformations B-C and D-A ranging from 0.1 MPa and 6 MPa are polytropic compressions with power given by:

$$P_{comp} = \frac{\dot{n}_{gas} L_{comp}}{\eta_{comp}} = \frac{\dot{n}_{gas}}{\eta_{comp}} \frac{m\,R\,T_{in,c}}{m-1} \left[1 - \left(\frac{p_{out}}{p_{in}} \right)^{\frac{m-1}{m}} \right] \tag{9.21}$$

where $\dot{n}_{gas}$ is gas flow, η_{comp} is the compressor efficiency, m is the polytropic coefficient, R is the universal gas constant, $T_{in,c}$ is the inlet gas temperature, p_{in} and p_{out} are the inlet and outlet compressor pressures. Compressors are powered directly by the bus-bar. The storage and the compression sub-systems data are detailed in Table 9.5.

Figure 9.16 shows the total power requested by the load in one year, while Figure 9.17 indicates the power obtained from the fuel cell, which directly feeds the load when solar radiation is high enough.

Figure 9.18 shows the quantity of hydrogen consumed by the fuel cell over a simulation period of one year.

The maximum amount of hydrogen produced is reached on day 335 at 15.8×10^3 mol, while the minimum storage is on day 46, when only 5.4×10^3 mol are totalled in the tank cluster (Figure 9.19). Since the initial charge is not completely consumed during the first months of the year, the hydrogen storage system can be scaled down to smaller or fewer tanks, which will be helpful in reducing the system costs in real-life applications. Smaller storage capacity though would entail hydrogen

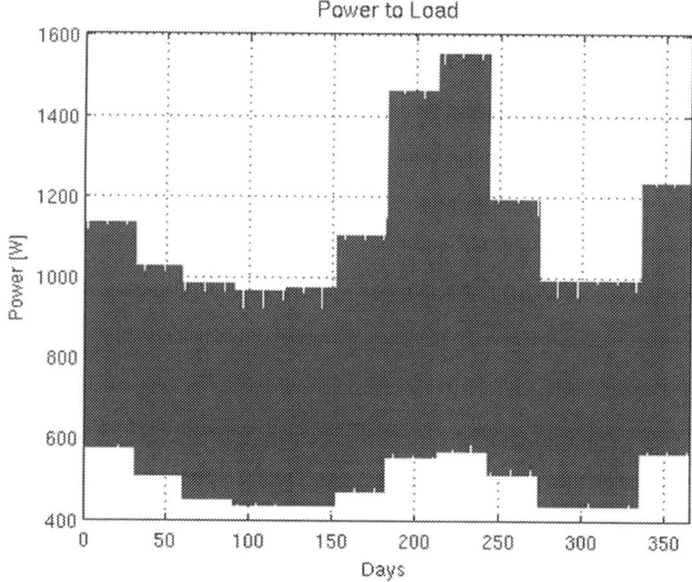

Fig. 9.16. Total power in one year (Reproduced with permission from [45])

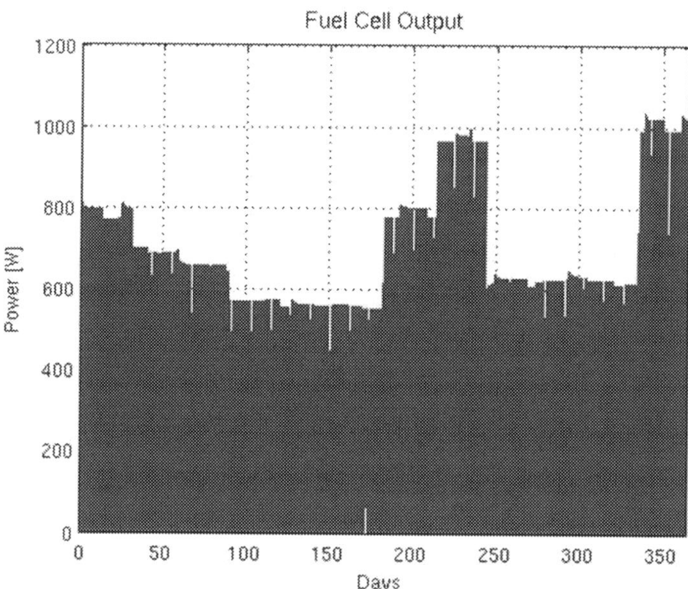

Fig. 9.17. Power from the fuel cell in one year (Reproduced with permission from [45])

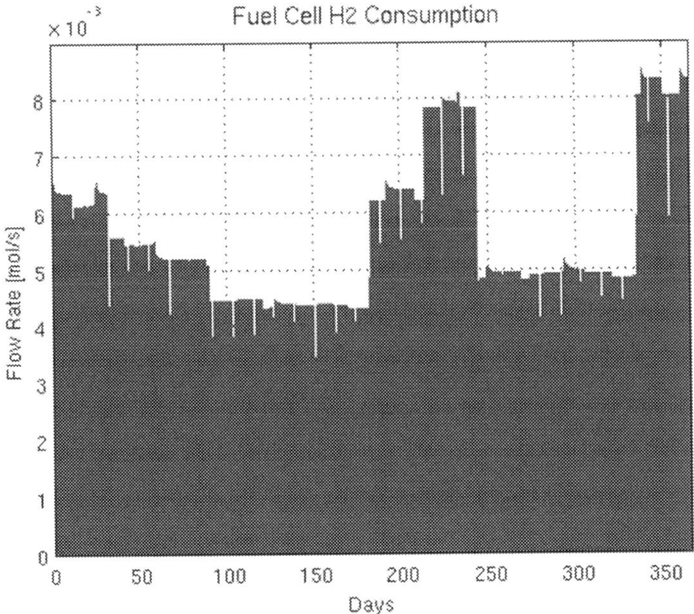

Fig. 9.18. Hydrogen consumption by the fuel cell in one year (Reproduced with permission from [45])

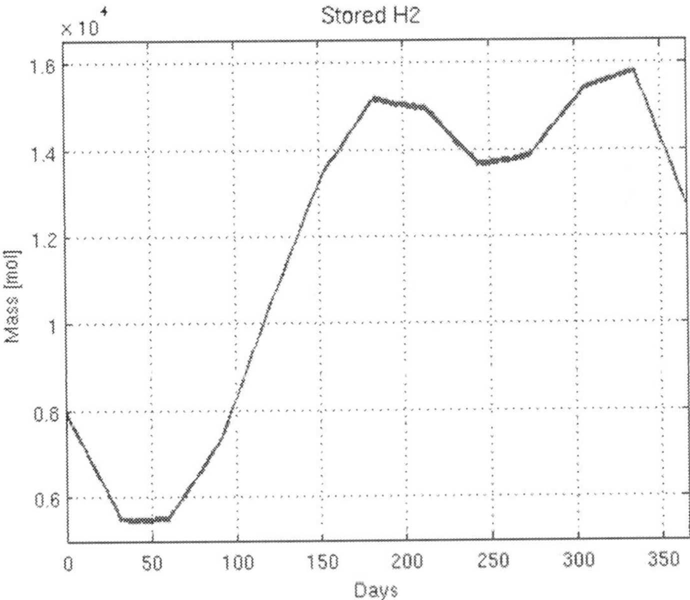

Fig. 9.19. Hydrogen mass stored over one year (Reproduced with permission from [45])

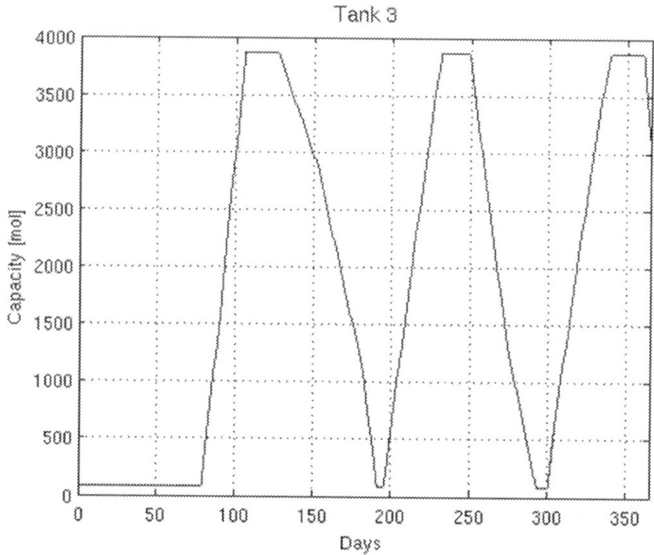

Fig. 9.20. Charging-discharging cycle capacity for tank 3 in one year (Reproduced with permission from [45])

venting during the summer months when hydrogen production is at its highest. In this system, no hydrogen is ever vented.

To examine the behaviour of the tank cluster, Figure 9.20 reports the charging and discharging cycles for tank 3 of the five tank. A complete charging cycle takes more than 20 days while discharging can take even longer. The time frame between the end and the start of discharging is around 33.6 h. With such a long operating cycle, it is possible to effectively vary the temperature of the AC bulk with, in principle, fairly straightforward temperature control systems. The total exchanged thermal energy during the charging-discharging cycles is as high as 2.246×10^9 J. The tanks are compact and produce a surplus at the end of the year, demonstrating the stand-alone capabilities of the system. All the sub-systems are suitable for stationary applications in terms of their dimensions and weight.

The surplus of hydrogen at the end of the year amounts to 4.8×10^3 mol, meaning that the system can operate as a stand-alone power unit. Some of the most relevant performance data are summarized in Table 9.6.

As already discussed, in the η_{Sys} estimation, no electric power to the satellite systems (like control logic, piping systems, etc.) has been taken into account, neither have the battery and the cluster initial storage. Furthermore, no thermal energy from the functioning of the electrolyser and the fuel cell is recovered, although combined heat and power (CHP) operations can be considered to increase the overall system efficiency.

Table 9.6. System performance data over a one-year period (Reproduced with permission from [45])

Total irradiation H_T	6.75 GJ/m^2
kWh$_{el}$/kW$_p$	1921
kg$_{H_2}$/kW$_p$	15.8
Shortest discharging-charging time (Tank 4)	33.6 h
η_{El}	66.9%
η_{FC}	63.0%
η_{DR}	90.7%
η_{HL}	36.3%
η_{Sys}	9.5%
Minimum stored H$_2$ (day 46)	5.4×10^3 mol
Maximum stored H$_2$ (day 335)	15.8×10^3 mol
H$_2$ one-year surplus	4.8×10^3 mol
Total exchanged thermal energy for storage cluster	2.246×10^9 J

9.6 Simulation with Wind Energy Conversion, Compression and Activated-Carbon Storage

The hydrogen energy system discussed in Section 9.4 is now simulated with wind energy as the only renewable energy source. Both the traditional compression and the activated-carbon storages are employed. The related wind energy modelling has been presented in Chapter 5.

Also in this case, all the systems are connected to a central bus-bar through buck, boost and DC-AC converters. A VRLA battery is employed to avoid fluctuations on the bus-bar and to supply the load until the fuel cell is turned on. The same control logic detailed in Section 9.2 is adopted to properly drive the various system inter-operations. Load data have been modelled as in Section 9.4.

The aerogenerator is a 3-blade wind turbine that converts the kinetic energy in the wind flow into electric energy. When the wind energy is sufficient, the load is directly powered by the aerogenerator. Otherwise, the batteries are discharged until they reach their lower SOC. Hydrogen then flows from storage to the fuel cell to recharge the batteries and to help supply the load. When there is a surplus of wind energy, the electrolyser is activated to restore hydrogen capacity in the storage tanks. The tanks can be either pressurised or filled with activated carbons in a powder form (as previously, AX-21).

From the experimental data, the wind speed profile in an certain location can be modelled after the Weibull's probability distribution function as given in:

$$h(v) = \left(\frac{k}{c}\right)\left(\frac{v}{c}\right)^{k-1} \exp\left(\frac{-v}{c}\right)^k \tag{9.22}$$

with the *form factor k* ranging between 1.5 and 2.5 and the *scale factor c* between 5 and 10 m/s. The form factor defines the shape of the function while the scale factor increases with the number of days characterized by higher wind speeds.

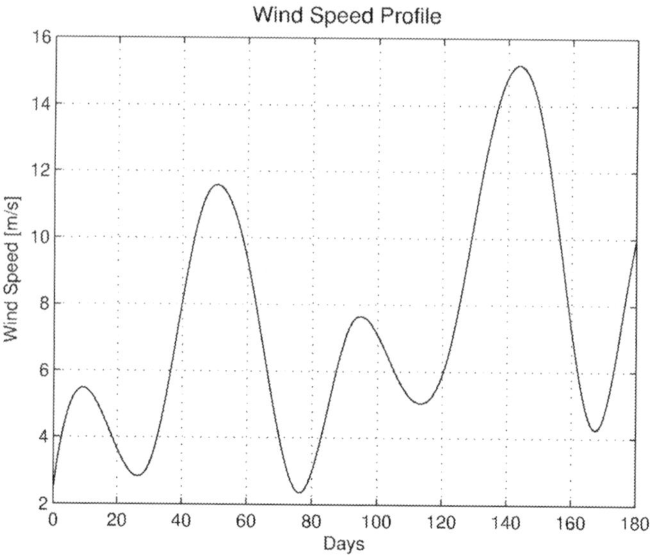

Fig. 9.21. Wind profile over a 6-month period (Reproduced with permission from [46])

Although the wind speed can vary significantly even within just minutes, to obtain a wind speed profile over time, a set of random points from the Weibull's probability distribution function has been generated and interpolated with splines to yield the curve in Figure 9.21. More realistic simulation results would be obtained by employing real wind data.

The power contained in a wind flow crossing a surface A with speed v is given by:

$$P_{wind} = \frac{\rho}{2} A v^3 = \frac{\rho}{2} \frac{D^2 \pi}{4} v^3 \qquad (9.23)$$

where D is the rotor diameter and ρ is the air density (determined by temperature, humidity, air pressure) that can be empirically described for sites at a height of 6000 m by the following equation:

$$\rho = \rho_0 \exp\left(\frac{-0.297}{3048} H_m\right) \qquad (9.24)$$

with ρ_0 as the air density at sea level and H_m as the site altitude.

As already discussed in Chapter 5, to account for the aerodynamic losses, P_{wind} must be multiplied by the power coefficient c_p, by the mechanical losses η_{mech}, by the asynchronous generator losses η_{asyn} and by the AC-DC conversion losses η_{ac-dc}. The DC electric power P_{aero} provided by the aerogenerator is then expressed as:

$$P_{aero} = \left(c_p \, \eta_{mech} \, \eta_{asyn} \, \eta_{ac-dc}\right) P_{wind}. \qquad (9.25)$$

To avoid rotation at ineffective low wind velocities, the rotor movement is not activated for wind speeds below the cut-in speed. Between the cut-in and the contin-

Table 9.7. Aerogenerator and activated carbon storage system data (Reproduced with permission from [46])

Aerogenerator	(3-blade)		
Diameter	5.5 m	Site elevation	50 m
ρ_0	1204 kg/m^3	Cut-in speed	5 m/s
Cut-out speed	22.5 m/s	Constant speed cut-out	15 m/s
c_p	0.45	Mech. losses	0.6
Asynch. gen. losses	0.92	AC-DC conv. losses	0.9
AC storage			
Initial hydrogen	1329 mol	AC type	AX-21
Specific surface	2745 m^2/g	Density	0.3 kg/l
Number of tanks	5	One tank storage	1231 mol
Tank volume	0.0764 m^3	Inner wall width	4 mm
Outer wall width	2 mm	Vacuum cavity w.	36 mm
Vacuum conductivity	0.004 W/mK	Radiation shields	20
Operating temp.	77–153 K	Cooling fluid	N
Cooling pipe volume	Not considered		

uous speed cut-out, a maximum power tracking control logic maximizes the power conversion by regulating the rotation speed to maintain c_p constant. To avoid turbine damages, an embedded control logic ensures that the rotor is limited to a constant speed above a pre-determined threshold speed, or it can stop altogether when the wind speeds are excessive.

The simulation spans over a 180 day period. Table 9.7 outlines the aerogenerator system data.

Figure 9.22 shows the aerogenerator electric power output. Notice the cut-ins (occurring at 5 m/s) and continuous speed cut-outs (at 15 m/s) that cause reductions in the power available to the load.

The power supplied to the electrolyzer is shown in Figure 9.23 for compression storage and in Figure 9.24 for compression storage.

The power obtained by the combustion of hydrogen in the fuel cell is depicted in Figure 9.25. By comparison with Figure 9.22, it is possible to notice that when wind energy is available, hydrogen is produced by the electrolyser, when the wind energy is not sufficient to supply the load, hydrogen is employed in the fuel cell.

The power supplied to the load is given in Figure 9.26.

The hydrogen capacity (Figure 9.27 for compression storage, Figure 9.28 for activated-carbon storage) varies over time according to the wind energy (compare with Figure 9.21). When the wind is low or the power to load overcomes the energy supply, the hydrogen capacity decreases, only to rise when the wind energy compensates the energy consumption more than sufficiently. The final capacity at the end of the period exceeds the initial capacity and indicates the stand-alone capability of the system.

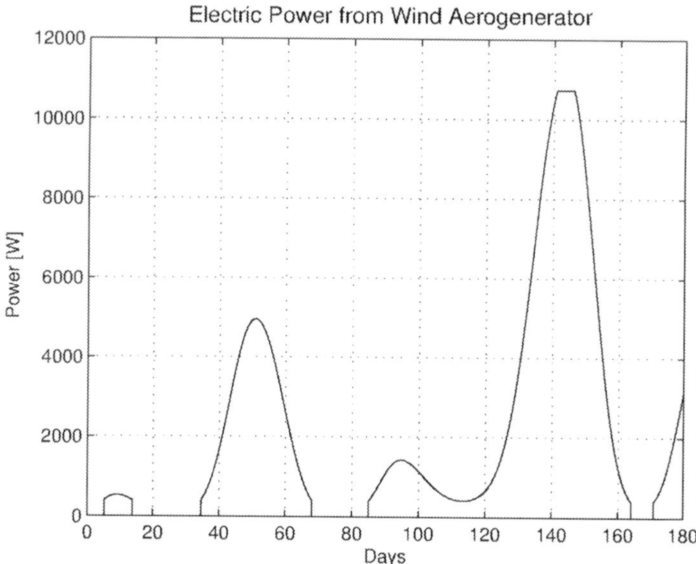

Fig. 9.22. Aerogenerator electric power output (Reproduced with permission from [46])

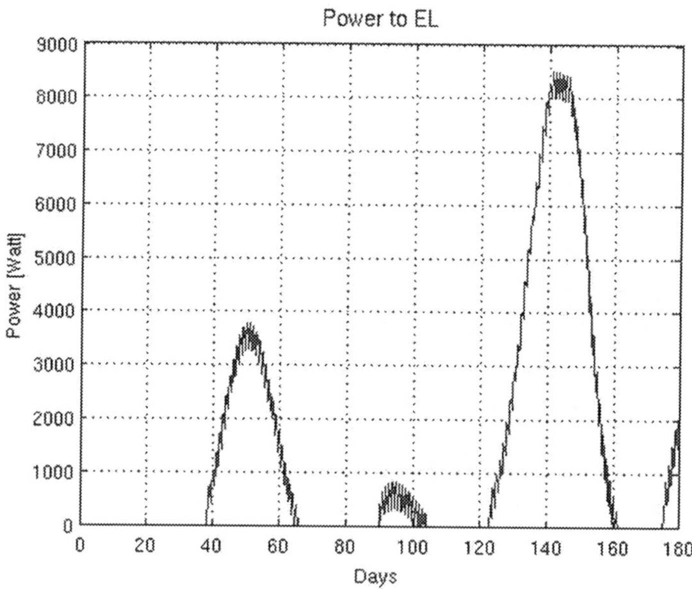

Fig. 9.23. Power to electrolyser, compression storage (Reproduced with permission from [46])

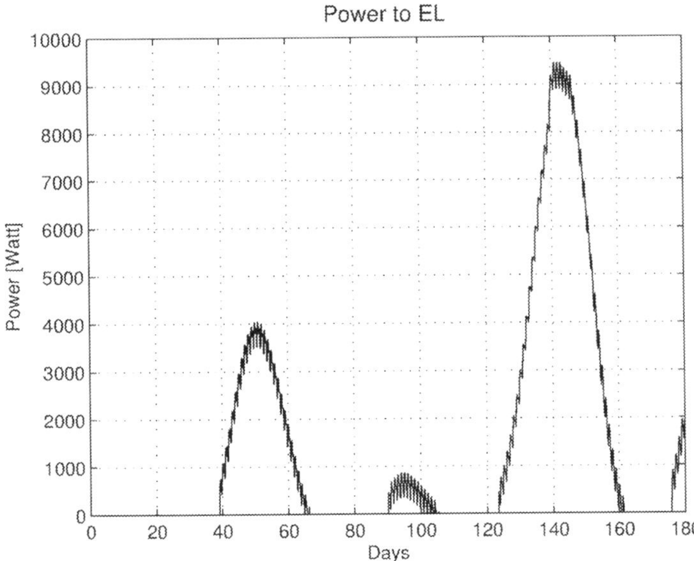

Fig. 9.24. Power to electrolyser, activated-carbon storage (Reproduced with permission from [46])

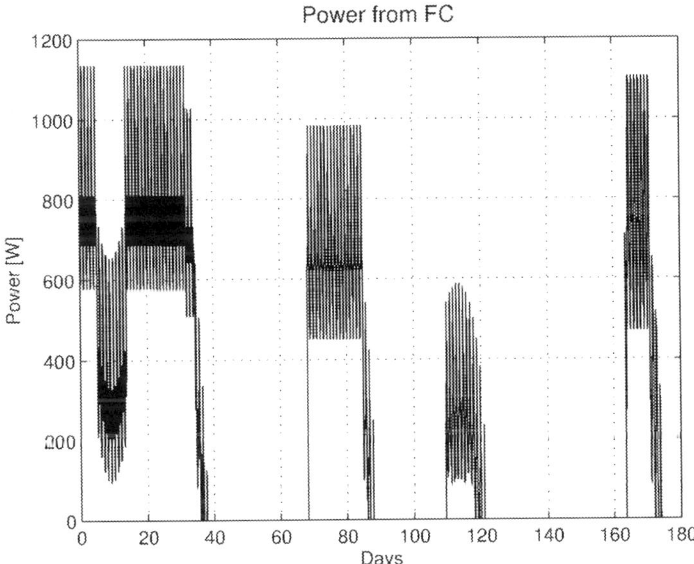

Fig. 9.25. Power from hydrogen consumption in fuel cell, activated-carbon storage (Reproduced with permission from [46])

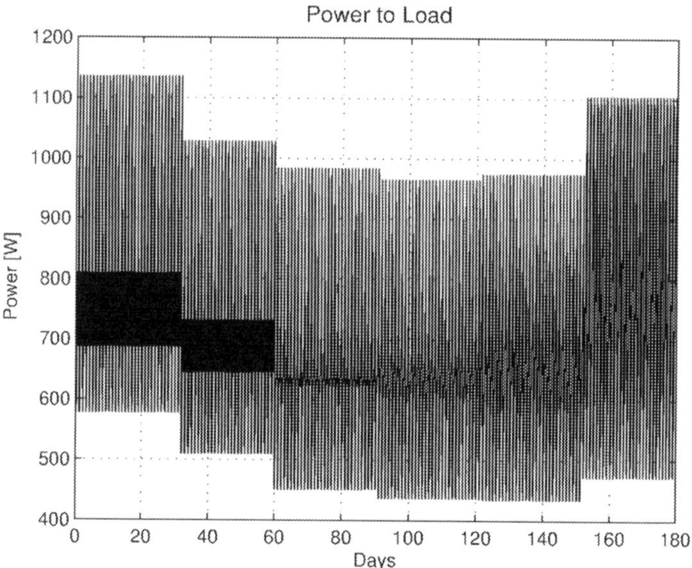

Fig. 9.26. Power supplied to load (Reproduced with permission from [46])

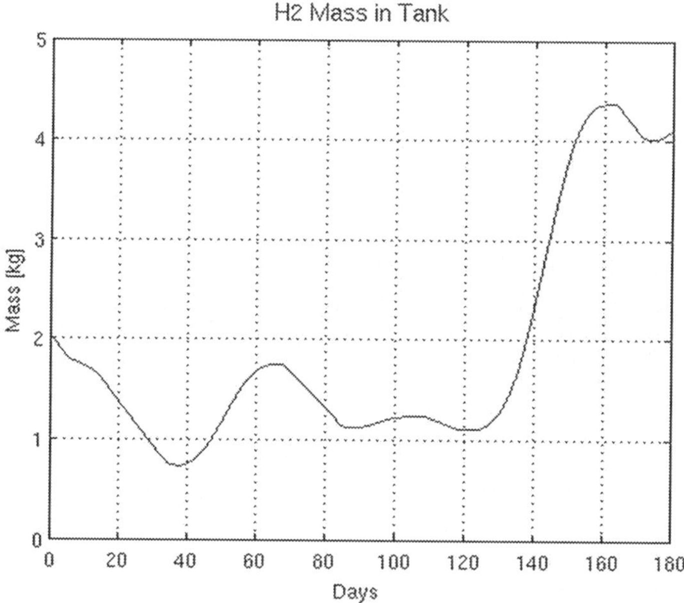

Fig. 9.27. Hydrogen capacity over a 6-month period, compression storage

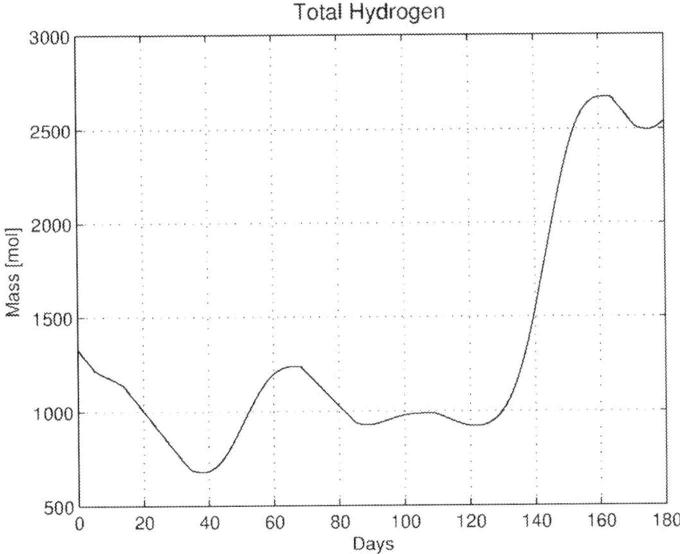

Fig. 9.28. Hydrogen capacity over a 6-month period, activated-carbon storage (Reproduced with permission from [46])

To compare physisorption storage with the classic storage technology, Table 9.8 summarizes the performance data deriving from the computations on both systems.

Table 9.8. Performance comparison of the wind-hydrogen energy systems (Reproduced with permission from [46])

Compression storage performance	
η_{Wind}	21.7%
η_{FC}	62.3%
η_{EL}	67.8%
η_{Sys}	12.7%
Minimum capacity (day 38)	0.734 g (364 mol)
Maximum capacity (day 163)	4366 g (2165 mol)
H_2 surplus at end of period	2087 g (1035.2 mol)

Physisorption storage performance	
η_{Wind}	21.7%
η_{FC}	62.3%
η_{EL}	69.2%
η_{Sys}	13.5%
Minimum capacity (day 38)	1366.6 g (677.9 mol)
Maximum capacity (day 161)	5380.7 g (2669 mol)
H_2 surplus at end of period	2388.9 g (1185 mol)

Each activated carbon tank has a volume of 0.0764 m^3, which makes for an overall volume of 0.382 m^3 (no inner cooling pipe volume is considered). These tanks are compact and can store a total of 6155 mol (12.4 kg) of hydrogen at 6 MPa, 77 K. The same amount of hydrogen stored by compression in a 0.382 m^3 tank would require a peak operating pressure of nearly 40 MPa at ambient temperature.

9.7 Notes on Exergy Analysis

The efficiencies calculated in the previous section are based on the first principle of thermodynamics. A complete energy analysis entails resorting to an exergy analysis as performed in [14] on the solar hydrogen energy system with photovoltaic conversion described in Section 9.5.

The system exergy efficiency is defined as the difference between exergy products and exergy inputs (excluding solar exergy input) divided by the solar exergy received by the system. The system exergy efficiency is estimated to be 4.0%, where most of the exergy destruction is performed by the PV sub-system, with an exergy efficiency of 14% when the rest of the plant reaches 29%.

To improve exergy efficiency, it is possible to recover all or part of the hot thermal energy, the cold thermal flows and generally all the products that contain exergy. This way, the exergy efficiency can rise to an 11% from the previous 4%. Another optimisation would be to adopt newer photovoltaic technologies with higher performances or to employ other renewable sources with higher first and second principle efficiencies.

9.8 Remarks on the Simulation of Solar Hydrogen Energy Systems

Solar hydrogen energy systems can be considered a feasible way for stand-alone energy management. Complete system efficiencies depend very much upon the kind of renewable energy source chosen. The advances in renewable energy source conversion technology, together with the optimisation of control algorithms and the retrieval of all the available exergy, cater for potential noteworthy improvements in η_{Sys}.

Storage is a critical part of the system and non-traditional storage technologies like carbon nano-structures or hydrides can be viable and safe substitutes for traditional technologies like compression or liquefaction.

Finally, hydrogen can also be employed as discussed in Section 2.4 without resorting to fuel cells. Such applications can also be an interesting way for hydrogen to function as an energy carrier either for stationary or non-stationary uses.

References

1. Ahmad G E, El Shenawy E T (2006) Optimized photovoltaic system for hydrogen production. Renewable Energy 31:1043–1054
2. Aiche-Hamane L, Belhamel M, Benyoucef B, Hamane M (2009) Feasibility study of hydrogen production from wind power in the region of Ghardaia. Int. J. Hydrogen Energy 34:4947–4952
3. Amendola S C, Sharp-Goldman S L, Saleem Janjua M et al (2000) A safe, portable, hydrogen gas generator using aqueous borohydride solution and Ru catalyst. Int. J. Hydrogen Energy 10 (25):969–975
4. Bilgen E (2001) Solar hydrogen from photovoltaic-electrolyzer systems. Energy Convers Manage 42:1047–1057
5. Bilgen E (2004) Domestic hydrogen production using renewable energy. Solar Energy 77:47–55
6. Bilodeau A, Agbossou K (2006) Control analysis of renewable energy system with hydrogen storage for residential applications. Journal of Power Sources 162:757–764
7. Bilgen C, Bilgen E (1984) An assessment on hydrogen production using central receiver solar systems. Int. J. Hydrogen Energy 9 (3):197–204
8. Deshmukh S S, Boehm R F (2008) Review of modeling details related to renewably powered hydrogen systems. Renewable and Sustainable Energy Review 12:2301–2330
9. El-Hefnawi S H (1998) Photovoltaic diesel-generator hybrid power system sizing. Renewable Energy 1 (11):33–40
10. El-Shatter Th F, Eskandar M N, El-Hagry M T (2002) Hybrid PV/fuel cell system design and simulation. Renewable Energy 27:479–485
11. Friberg R (1993) A photovoltaic solar-hydrogen power plant for rural electrification in India, Part 1: a general survey of technologies applicable within the solar-hydrogen concept. Int. J. Hydrogen Energy 18 (10):853–882
12. Goldstein L H, Case G R (1978) PVSS-A photovoltaic system simulation program. Solar Energy 21:37–43
13. Griesshaber W, Sick F (1991) Simulation of Hydrogen-Oxygen Systems with PV for the Self-Sufficient Solar House (in German). FhG–ISE, Freiburg im Breisgau, Germany
14. Hacatoglu K, Dincer I, Rosen M A (2011) Exergy analysis of a hybrid solar hydrogen system with activated carbon storage, Int. J. Hydrogen Energy 36:3273–3282
15. Hammache A, Bilgen E (1987) Photovoltaic hydrogen production for remote communities in Northern latitudes. Solar & Wind Technology 2 (4):139–144
16. Havre K, Gaudernack B, Alm L K, Nygaard T A (1993) Stand-Alone Power Systems based on renewable energy sources. Report no. IFE/KR/F-93/141, Institute for Energy Technology, Kjeller, Norway
17 Hollenberg W, Chen E N, Lakeram K, Modroukas D (1995) Development of a photovoltaic energy conversion system with hydrogen energy storage. Int. J. Hydrogen Energy 3 (20):239–243
18. Hug W, Divisek J, Mergel J, Seeger W et al (1992) Highly efficient advanced alkaline electrolyzer for solar operation. Int. J. Hydrogen Energy 17 (9):699–705.
19. Kélouwani S, Agbossou K, Chahine R (2005) Model for energy conversion in renewable energy system with hydrogen storage. Journal of Power Sources 140:392–399
20. Kennerud K L (1969) A technique for identifying the cause of performance degradation in cadmium sulfide solar cells. 4th IECEC 561–566
21. Khan M J, Iqbal M T (2005) Dynamic modeling and simulation of a small wind-fuel cell hybrid energy system. Renewable Energy 30:421–439

22. Khan M J, Iqbal M T (2009) Analysis of a small wind-hydrogen stand-alone hybrid energy system. Applied Energy 86:2429–2442
23. Kolhe M, Agbossou K, Hamelin J, Bose T K (2003) Analytical model for predicting the performance of photovoltaic array coupled with a wind turbine in a stand-alone renewable energy system based on hydrogen. Renewable Energy 28:727–742
24. Lodhi M A K (1997) Photovoltaics and hydrogen: future energy options. Energy Convers Manage 18 (38):1881-1893
25. Maclay J D, Brouwer J, Scott Samuelsen G (2007) Dynamic modeling of hybrid energy storage systems coupled to photovoltaic generation in residential applications. J. Power Sources 163:916–925
26. Mantz R J, De Battista H (2008) Hydrogen production from idle generation capacity of wind turbines. Int. J. Hydrogen Energy 33:4291–4300
27. Milani M, Montorsi L, Golovitchev V (2008). Combined Hydrogen Heat Steam and Power Generation System. Proc. 16th ISTVS Int. Conference, Turin, November 25–28
28. Muselli M, Notton G, Louche A (1999) Design of hybrid-photovoltaic power generator, with optimization of Energy Management. Solar Energy 3 (65):143–57
29. Pedrazzi S, Zini G, Tartarini P (2010) Complete modeling and software implementation of a virtual solar hydrogen hybrid system. Energy Conversion and Management 51 (1):122–129
30. Sherif S A, Barbir F, Veziroğlu T N (2005) Wind energy and the hydrogen economy – review of the technology. Solar Energy 78:647–660
31. Siegel R, Howell J R (2002) Thermal Radiation Heat Transfer, Hemisphere, New York
32. Sopian K, Ibrahim M Z, Wan Daud W R, Othman M Y et al (2009) Performance of a PV-wind hybrid system for hydrogen production. Renewable Energy 34:1973–1978
33. Sternfeld H J, Heinrich P (1989) A demonstration plant for the hydrogen/oxygen spinning reserve. International Journal of Hydrogen Energy 14:703–716
34. Sürgevil T, Akpinar E (2005) Modelling of a 5 kW wind energy conversion system with induction generator and comparison with experimental results. Renewable Energy 30:913–929
35. Ulleberg Ø, Morner S O (1997) Trnsys simulation models for solar hydrogen systems. Solar Energy 4-6 (59):271–279
36. Vemulapalli G K (1993) Physical Chemistry. Prentice Hall, New Delhi
37. Zhang J, Fisher T S, Veeraraghavan Ramanchandran P, Gore J P et al (2005) A review of heat transfer issues in hydrogen storage technologies. J Heat Transfer 127:1391–1399
38. Zhou L, Zhang J (2001) A simple isotherm equation for modeling the adsorption equilibria on porous solids over wide temperature ranges. Langmuir 17:5503–5507
39. Zhou L, Zhou Y, Sun Y (2004) Enhanced storage of hydrogen at the temperature of liquid nitrogen. Int. J. Hydrogen Energy 29:319–322
40. Zhou L (2005) Progress and problems in hydrogen storage methods. Renewable and Sustainable Energy Reviews 9:395–408
41. Zhou L, Zhou Y, Sun Y (2006) Studies on the mechanism and capacity of hydrogen uptake by physisorption-based materials. Int. J. of Hydrogen Energy 31:259–264
42. Zhou L, Zhou Y (1998) Linearization of adsorption isotherms for high-pressure applications. Chem. Eng. Sci. 14 (53):2531–2536
43. Zini G, Tartarini P (2009) Hybrid systems for solar hydrogen: a selection of case-studies. Applied Thermal Engineering 29:2585–2595
44. Zini G, Marazzi R, Pedrazzi S, Tartarini P (2009) Solar hydrogen hybrid system with carbon storage, International Conference on Hydrogen Production ICH2P-09, Toronto, 2–6 May

45. Zini G, Tartarini P (2010) A solar hydrogen hybrid system with activated carbon storage. International Journal of Hydrogen Energy 35 (10):4909–4917
46. Zini G, Tartarini P (2010) Wind-hydrogen energy stand-alone system with carbon storage: Modeling and simulation. Renewable Energy 35 (11):2461–2467
47. Züttel A, Nützenadel Ch, Sudan P, Mauron Ph et al (2002). Hydrogen sorption by carbon nanotubes and other carbon nanostructures. Journal of Alloys and Compounds (330-332):676–682
48. Züttel A, Wenger P, Rentsch S, Sudan P et al (2003) LiBH4 a new hydrogen storage material. J. Power Sources 1-2 (118):1–7

10

Real-Life Implementations of Solar Hydrogen Energy Systems

Real-life implementations of solar hydrogen energy systems, both for stationary and non-stationary applications, are becoming ever more common with new technology advances. This chapter examines some of the most interesting actual installations, in order to highlight the peculiarities and the benefits of these systems and to bring to light the shortcomings to deal with, so that the readers can probe into the practical side of these solar hydrogen solutions.

10.1 Introduction

To-date the actual implementations of the complete solar hydrogen energy systems as demonstrated in Figure 9.1 are still not very wide-spread. The different combinations of these systems have been covered in the related literature. The following sections will briefly describe some of the most interesting solutions and the reader can also find relevant references listed at the end of the chapter.

10.2 The FIRST Project

A stand-alone system converting solar energy directly in hydrogen has been implemented and studied at CIEMAT (Madrid, Spain) with the project *FIRST - Fuel cell Innovative Remote System for Telecoms* [2, 3]. This stand-alone solution for remote telecom stations consists of a 1.4 kWkW$_p$ CIS PV plant that supplies 200 W with a conversion efficiency of nearly 10%. The storage employs batteries (20 kWh in 24 lead-acid units) to smooth the PV yield variations. A PEM electrolyser produces hydrogen at a pressure of 30 bar for medium and long-term storage into 7 metallic hydride tanks; each of them has a capacity of 10 Nm^3 with charge-discharge cycles ranging between 0 to 40 °C.

A PEM, dead-end mode fuel cell converts hydrogen when needed, with a net power of 75–280 W capable of supplying the load for an entire month. A control

Zini G., Tartarini P.: Solar Hydrogen Energy Systems. Science and Technology for the Hydrogen Economy. DOI 10.1007/978-88-470-1998-0_10, © Springer-Verlag Italia 2012

software drives the system operations and a data-logger registers data like conversion efficiencies, battery SOC, the electric current yielded by the system or required by the load, operating temperatures and storage pressures.

The incident solar radiation has been measured in the range of 82 kWh, while the PV field has converted an energy of nearly 5 kWh. The electrolyser has been supplied with 1.5 kWh for a total production of 0.33 Nm^3 (equivalent to 1 kWh at LHV). The load has required 3.4 kWh with a 0.5 kWh drained by the energy management system. The energy converted by the PV field maintains an average SOC of 87% and a daily hydrogen flow capable of completely supplying the load with a 2.5 A per day. During the summer, the energy converted by the PV field is sufficient to power up the load (2.7 kWh/day) and the electrolyser without the need to switch the fuel cell on. The daily hydrogen production of 0.33 Nm^3 is sufficient to have an overcapacity at the end of spring-summer season capable of granting one month's operations with no assistance from the PV field. If no long-term technical issues arise, the system can be fully independent to supply energy to the load.

The electrolyser is switched on mostly during daytime when solar radiation is high. The storage pressure increases linearly from the minimum to the maximum approximately in 5 hours. This length of time needs to take into account the slow adsorption kinetic of the hydrides chosen for storage. A way to decrease such amount of time would then be to increase the number of hydride storing units.

The fuel cell is activated when batteries reach an SOC of 35%; at this point, the fuel cell supplies the load with 280 W until SOC returns to values over 70%. The fuel cell is then turned off and batteries can be switched on again when the PV field output energy is not enough to independently supply the load.

The conversion efficiency is computed as:

$$\eta_T = \frac{E_C}{E_S} = \frac{\eta_{PV}\,E_{PV} + \eta_H\,E_H}{E_{PV} + E_H + E_0} \tag{10.1}$$

where E_C is the energy supplied to the load, E_S is the total energy from the incident solar radiation, η_{PV} is the path 1 (PV $\rightarrow$ Battery $\rightarrow$ Load) conversion efficiency, η_H is the path 2 (PV $\rightarrow$ Battery $\rightarrow$ Electrolyser $\rightarrow$ Storage $\rightarrow$ FuelCell $\rightarrow$ Battery $\rightarrow$ Load) conversion efficiency, E_{PV} is the path 1 input energy, E_H is the path 2 input energy and finally E_0 is the input energy which is not converted.

The system management software must minimize E_0 by increasing the energy that enters paths 1 and 2. η_T has reached, in test conditions, values around 6–7%, while η_H has been computed as 2.3%. Using a hydrogen storage system grants the possibility to increase the overall performance of the system. Indeed, all the energy available from the sun starting from 12 am is lost if the electrochemical batteries have already reached the maximum SOC. The overall system efficiency is then much lower (around 4%) unless the electrochemical storage is sized to accept all incoming energy.

10.3 The Schatz Solar Hydrogen Project

Started in 1989 at Humboldt State University Telonicher Marine Laboratory (California, USA), the project has the goal of verifying the feasibility of solar hydrogen in a real operating context, in this case the aquarium aeration [7]. Equipped with traditional compression storage, the system receives energy from a PV field with a peak total power of 9.2 kW_p. The electrolyser belongs to a medium pressure alkaline type comprising of 12 cells and capable of yielding 20 l/min (at standard conditions) at 240 A, 24 V, direct current. Hydrogen is supplied at a pressure of 790 kPa and stored in 3 vessels for a total capacity of 5.7 m^3. The fuel cell can convert up to 133 kWh (considering hydrogen HHV) and supply a load at 600 W for 110 h with an efficiency of 50%.

When possible, the electric energy obtained from the photovoltaic field is directly routed to the load, while every power surplus is used to generate hydrogen. In case solar radiation is insufficient to provide enough power, hydrogen is employed in the fuel cell to be converted into electric energy to make up for the power shortage. The overall system is controlled automatically by an unsupervised control logic.

The efficiencies are computed with the following relations:

- *Faraday's efficiency*: actual amount of H_2 production over theoretical H_2 production;
- *EL efficiency*: rate of H_2 production over EL input power;
- *PV efficiency*: PV power used by the system over solar irradiation;
- *H_2 production efficiency*: PV efficiency times EL efficiency.

The Faraday's efficiency measured for the system is as high as 94%, while the electroyser efficiency is around 76% with a final overall system efficiency ranging around 6%. The average daily efficiencies are characterised by acceptable values during a period of 11 months, representative of a wide range of solar radiation variability. The electrolyser efficiency is above 75% for more than 70% of the daily averages, with the current densities variable from 15 to 400 mA/cm^2 and the electrolyte temperatures from 10 to 70 °C. The photovoltaic field efficiency surpasses 7.8% for 70% of the daily averages, while the hydrogen production efficiency is over 6% for the same percentage of daily averages.

The control system is able to ensure that the load is supplied in a uniform and stable way also during days with unfavourable meteorological conditions and with very uneven solar radiation averages. Apart from a few minor issues related to some of the photovoltaic modules, the continuity and the quality of the service have been proven to be remarkable.

Since the plant has been constructed and operated in real working conditions, all necessary precautions have been taken to ensure a safe operation. The system design has taken into account a thorough risk analysis to avoid minor to catastrophic failures. Fault and safety conditions have been studied with a fault tree analysis technique, taking advantage of field tests and advanced debugging techniques. The personnel have been trained in the management and operations of chemical gases and fluids so that the emergency procedures can be implemented flawlessly. A fail-safe system has been installed to automatically switch off the system in the case of leakage or other predetermined fault conditions. Finally, maintenance plans have been put in place to assure a periodical assessment of the system integrity and of the full operating capabilities.

In this way, the plant has been operated with very minor issues, showing how professionalism in hydrogen management is the key to boost productivity and to guarantee safety.

10.4 The ENEA Project

The Italian *ENEA* organisation has devised a research project to understand the technical aspects of hydrogen storage from solar energy photovoltaic conversion [4]. The main goals of the project were to evaluate the long-term reliability of the system and the issues related to safe operations and maintenance.

The set-up comprises of a photovoltaic field of 5.6 kW_p (180 modules, monocrystalline), a 5 kW bipolar alkaline electrolyser (water-cooled, 80 °C, high-pressure, 20 bar output), a mixed metal hydride and pressure tank storage sub-system (with a storage capacity of 18 Nm^3) and a 3 kW PEM fuel cell operating at 72 °C.

The control system is based on a Programmable Logic Controller (PLC) and controls electrolyte temperatures, current ranges, water conductivity, pressures, gas leakages and impurities. It manages system on/off cycles and stops the system depending on the type of alarm.

In an effort to improve overall system efficiency, the storage system releases hydrogen at 20 °C between 2 and 4 bar, avoiding the need to supply heat to the downstream fuel cell. Hydrogen and air are supplied at constant pressure and are consumed evenly according to the varying load conditions. Water is either retrieved for system cooling and moistening of process gases, or vented in atmosphere. Part of the heat produced during operations is recovered in a water-air heat exchanger.

All the satellite and control sub-systems request around one tenth of the power produced by the overall system output, which is 2.75 kW at 125 A, 22 V. At maximum operating conditions, the fuel efficiency at hydrogen LHV is 53,6%.

The plant has been designed to properly account for all the safety issues related to the management of inflammable gases: the walls and the doors are fireproof, the electric wirings and the components are specifically-constructed for environments with high risk of explosion and the sensors have been installed to immediately signal emergency conditions.

The operation results confirm that the system is easy to operate and delivers good performances.

10.5 The Zollbruck Full Domestic System

A domestic system has been set up in Zollbruck, Switzerland and has been operating manually since 1991 by the home owners themselves [6].

The renewable energy source is solar radiation with a photovoltaic field of 5 kW_p and an average efficiency of 8.4%. The roof area covered is 65 m^2. The other subsystems are located in two 10 m^2 rooms and they include:

- a DC-DC converter (95% efficiency);
- an alkaline electrolyser of 5 kW (62% average efficiency);
- a H_2 purifying unit;
- a compressor;
- two metal hydride tanks, one of which is 15 Nm^3 for household appliances (a oven and a washing machine) and the other one is 16 Nm^3 used to supply power to a mini-van running on H_2.

Prior to storage in the metal hydride tanks, a compressor is used to store H_2 in a pressurized tank. The storage capacity is 10 times less than what would be necessary to achieve a complete energy independence from the grid, which can be reached if the storage capacity is increased to at least 200 Nm^3. In summer days, more than half of the H_2 production cannot be stored due to the absence of a temperature control system that would otherwise drive the adsorption kinetics inside the metal hydride tanks, so that all the available storage volume can be employed.

Another sub-optimality is caused by the loss of thermal energy which is not retrieved and used altogether.

A connection to the grid is of course needed to supply all the loads that are not directly connected to the hydrogen system.

An energetic balancing of the system shows that the 293 GJ received in one year by the PV field would become:

- 24.5 GJ electric energy before DC-DC conversion or 23.4 GJ after DC-DC conversion. Of this amount only 18.6 GJ are used by the electrolyser while the remaining 4.9 GJ are sent to the grid or used to charge auxiliary batteries (3 GJ per year);
- 11.5 GJ as H_2 stored (1148 Nm^3) with a 62% electrolyser efficiency; from which only 10.6 GJ remain after H_2 treatment and purification.

The net annual hydrogen production amounts to 16 Nm^3/m^2.

To improve the overall system efficiency, an automatic control system is to be added to manage the functioning and to avoid the implicit inconsistencies in manual operations. A treatment aiming to avoid H_2 losses would increase the quantity available for storage by 8%. Furthermore, increasing the storage capacity to 200 Nm^3 will make the system completely independent from the grid. Finally, the recovery of wasted thermal energy could also improve the system performance.

This type of hydrogen system has proven to be a straightforward implementation in domestic settings despite its room for further technical improvements.

10.6 The GlasHusEtt Project

The *GlasHusEtt* is a building in Stockholm, Sweden, used as an exposition and information center to inform citizens on sustainable life-styles and renewable energy technologies. The building has been sponsored and financed by Swedish institutions and industries in the sector.

Together with the traditional heat and electric installations, a solar hydrogen system has been set-up to allow the researchers to study the different performances and peculiarities by direct comparison [5]. Such system, which is the first installed in a residential setting in Sweden, integrates a photovoltaic plant, a biogas system, an electrolyser and a fuel cell.

The 3 kW_p, multi-crystalline PV field covers a roof area of 25 m^2. The fuel cell can use hydrogen from a reformer or directly from the tank storage. A water tank is used as a heat accumulator to recover the heat from the fuel cell. The electrolyser is connected to both the PV field and the electric grid, so that it can work both in DC and AC modes. The hydrogen production is placed under the roof where the 50 l storage tanks are directly connected to the electrolyser. The system is monitored by a control unit and a data-logger that records H_2 mass flow, electricity, the heat either produced or used by the different sub-systems and the meteorological data received from the sensors installed over the building roof.

The fuel cell has delivered an electric efficiency of 13% and a thermal efficiency of 56%, while the electrolyser efficiency is around 43%. Approximately 5% of the input energy is assumed to be lost in the exhaust, while the heat and power losses amount to 25.6%.

Simulations show that the system is not capable of providing stand-alone capabilities. The stored hydrogen is emptied in only 4 h while the filling takes 34 h to complete. The biogas heater cycles frequently from on to off working mode, but a lot of heat is lost in the environment. The heat storage is too small (500 l of water) to act as a heat sink, hence most of the heat from the fuel cell is dispersed in the room.

The overall system efficiency can be improved by a correctly dimensioned heat recovery system, a more powerful PV plant and a better hydrogen storage system.

10.7 The Trois Rivière Plant

The stand-alone system designed by the University of Québec Trois-Rivière employs both wind and sunlight to provide electric power to telecommunications stations [1]. The surplus energy is employed to produce and store hydrogen to be used when the two renewable energy sources are not sufficient to provide the load with the requested energy.

The system consists of:

- a 10 kW wind aero-generator with a 30 m tower;
- a 1 kW_p PV field;
- a set of batteries acting as a power buffer to a DC bus-bar;

- a 5 kW electrolyser producing 1 Nm^3/h of purified hydrogen at 7 bar;
- a compressor at 10 bar hydrogen pressure;
- a 3.8 m^3 tank that can store up to 125 kWh;
- a 5 kW PEM fuel cell;
- a DC/AC inverter to supply 115 V at 60 Hz to the load.

The hydrogen production starts when the wind speed hits 3.4 m/s and reaches the maximum of 10 kW with a wind speed of 13 m/s. At an average speed of 6 m/s, the aerogenerator provides an average power of 2 kW and a hydrogen production rate of 0.4 Nm^3/h. The output voltage from the wind and the PV is converted to 48 V on a DC bus-bar, whose voltage is stabilized by means of batteries. The bus-bar is connected to the aerogenerator, the PV field, the fuel cell, the electrolyser, the compressor and the load.

The efficiency is computed as the ratio of the measured output power of the fuel cell and its theoretical power, as defined by:

$$P_{theoretical} = \frac{HHV \times N_{cells} \times I}{2e^- \times F} \tag{10.2}$$

where Ncells = 35 (stack number of cells), I is the load current and F is the Faraday's constant.

The electrolyser efficiency is as high as 65% at 23 °C and 71% at 55 °C. The fuel cell efficiency is higher than 45% at 4 kW, with a total efficiency over 42%.

The system has been confirmed to possess a very quick response capability to the changes in the load.

10.8 The SWB Industrial Plant

An industrial plant has been built in Neunburg vorm Wald (Germany) by Solar Wasserstoff Bayern GmbH (SWB), a joint venture of Bayernwerk AG, Siemens AG, BMW AG and Linde AG [10]. The project intends to figure out the best design, integration and operation strategy of several different technologies aiming to achieve efficient and stable energy yield from renewable energy sources in a dedicated industrial plant.

Many different technologies have been integrated and implemented in the power plant:

- photovoltaic technologies like mono, multi and amorphous silicon modules are working in parallel in different sizes ranging from 6 to 135 kW$_p$;
- DC/DC and DC/AC converters, DC and AC bus-bars;
- two low pressure electrolysers of 111 kW and 100 kW capable of hydrogen productivities of 47 m^3/h;
- compressors and various gas treatment stages;
- heaters (running on mixed gas and hydrogen) and refrigerating units (running on hydrogen) of 16.6 kW;

- three fuel cells: the first one is alkaline of 6.5 kW electric and 42.2 kW thermal, the second one is phosphoric acid of 79.3 kW electric and 13.3 kW thermal, and the third one is PEM at 10 kW which uses air to power an hydrogen-operated forklift;
- a filling station for hydrogen-powered cars.

In this plant, safety specifications and risk management have been regarded as the main project liabilities and concerns. Prevention measures were the key to the safety design. To avoid every possible danger coming from the explosive mixtures of hydrogen and air or oxygen, gas confinement and regulation have followed state-of-the-art guidelines, with appropriate construction and plant layout and adoption of venting, alarm and leakage control systems. Furthermore, to eliminate every potential detonation trigger, the use of free flames is abolished in all the compound. Also, anti-sparking components have been adopted in electric and mechanical systems; safety distances were created between different building areas and all the machines, motors, power systems and electric wirings have been earthed. Fire-resistant walls and doors and automatic fire extinguishing devices were installed and placed in areas with higher risks. Finally, emergency procedures, fire-fighting and evacuation plans, together with detailed maintenance plans and periodical checks of all safety systems were put in place and the personnel trained to perform all routine safety tasks as flawlessly as possible.

The management practices from the Neunburg vorm Wald plant have demonstrated that the existing knowledge on process engineering, safety design and plant operation techniques is valid and sufficiently advanced to run large-scale hydrogen power plants.

10.9 The HaRI Project

The system, developed under the *Hydrogen and Renewable Integration* project in Loughborough UK, has been built to provide electric and thermal energy to residential and commercial buildings [8] [9]. This implementation converts energy from the following sub-systems, making this an interesting example to examine the complex integration of different technologies:

- two 25 kW wind two-bladed turbines with stall control and squirrel-cage electric asynchronous generators;
- two 3 kW_p photovoltaic fields, one mono-crystalline and the other poly-crystalline;
- two hydroelectric turbines with respective capacities of 1 kW and 2.7 kW;
- a CHP system with internal combustion engine which delivers an electric capacity of 15 kW and a thermal capacity of 38 kW.

The electrolyser belongs to a high-pressure (25 bar) alkaline type with a yield of 8 Nm^3 per hour and a power of 34 kW. Its output flow can be controlled but only up to a limit of 20% to avoid inner cells contaminations. Since the continuous on/off cycles can increase the chance of cathode damages, a set of electrochemical traditional batteries are installed to reduce the effect of oscillations mainly caused by the aero-generator on the bus-bar. These batteries are characterised by a high-density charge (1000 Wh/kg) and high-temperature operations (300 °C), with a capacity of 32 Ah and 20 hours of stand-alone capability.

The hydrogen tanks have a 22.8 m^3 capacity and can operate up to 137 bar, for a maximum storage of 2856 Nm^3. To achieve those pressures, a 4 kW compressor is adopted.

The two fuel cells are PEM; one is devised for CHP (2 kW electric at 24 V and 2 kW thermal) and the other can produce 5 KW electric at 48 V, though not for CHP. The system has a central bus-bar backbone that connects all the sub-systems with a central control logic to oversee and optimise overall operations.

As already seen, a critical issue in this type of highly integrated systems is the need to find a trade-off between the different sub-system optimisations. Such compromises oblige some sub-systems to work in sub-optimal conditions. A good control logic therefore is required to reduce these sub-optimalities and to boost the overall system efficiency. This can also decrease the number of maintenance needed and prolong the system life-span. The most important sub-system to be controlled is reported to be the electrolyser, whose limited number of on/off cycles causes more energy loss due to the minimum power level at which it operates.

10.10 Results from Real-Life Implementations

The above real-life projects have provided many useful pieces of information as elaborated below.

First of all, the system performance and the hydrogen costs are correlated to the energy that can be converted from the renewable energy source. Reaching the grid-parity of photovoltaic or wind systems will be a prominent goal to achieve for solar hydrogen production to become competitive and sustainable. Advances in conversion technology will lead to much higher overall system efficiencies, improve the economic returns and facilitate market introduction. The cost evaluation should also consider the fact that the tank and the piping must use materials able to resist degradation phenomena like hydrogen embrittlement. A correct and thorough life-cycle analysis will then be crucial in understanding the economic viability.

Secondly, even though the energy conversion through a PV-EL-Storage-FC chain is more complex than a direct supply to the load, it can reduce or avoid energy losses by means of storage. A proper control logic and energy management system is then vital to obtain an optimal overall system performance. Furthermore, all the components, namely the use of electrolysis combined with renewable energy sources, have been proven to be extremely reliable.

With respect to the safety management, the importance of the satellite systems that ensure a secure and reliable operation cannot be stressed enough, regardless of the overall dimension or the type of integration of the system. A rigorous risk analysis must be performed to maximise safety and minimise all potential dangers. These hydrogen systems in fact can function at a very high safety level if correctly designed and carefully maintained by qualified professionals.

As for the energy supply fluctuations from the renewable sources, this volatility can be efficiently smoothed out by the downstream hydrogen system, which will also help achieve a target of 6000 operating hours per year. Some grid operators accept a maximum percentage of 10 to 20 of the overall production comprised by uneven energy sources and the adoption of hydrogen long-term storage will render the solar hydrogen systems more appealing to these grid operators, since the unpredictability of the intermittent sources no longer affects the output.

The examples listed in this chapters have confirmed that it is very straightforward to combine sources like photovoltaic, biogas, wind, hydroelectricity together with hydrogen systems. This means that different countries, according to their respective contexts, can opt for the best combination from a wide range of sources available. For instance, northern countries with less direct solar radiation will find wind energy or hydroelectricity preferable instead of photovoltaic conversion.

Finally, hydrogen systems are able to respond quickly and efficiently to the changes in load conditions. This is very important when the loads are characterised by frequent variations in electricity demand. Big hydrogen power systems would perform well in meeting these changing demands from the grid, as opposed to other traditional power plants that spend hours of ramp-up to start injecting energy into the grid.

References

1. Agbossou K, Chahine R, Hamelin J, Laurencelle F et al (2001) Renewable energy systems based on hydrogen for remote applications. Journal of Power Sources 96:168–172

2. Chaparro A M, Soler J, Escudero M J, Daza L (2003) Testing an isolated system powered by solar energy and PEM fuel cell with hydrogen generation. Fuel Cells Bull. 10–12

3. Chaparro A M, Soler J, Escudero M J, de Ceballos E M L et al (2005) Data results and operational experience with a solar hydrogen system. Journal of Power Sources 144:165–169

4. Galli S, Stefanoni M (1997) Development of a solar-hydrogen cycle in Italy. Int. J. Hydrogen Energy 5 (22):453–458

5. Hedström L, Wallmark C, Alvfors P, Rissanen M et al (2004) Description and modelling of the solar-hydrogen-biogas-fuel cell system in GlashusEtt. Journal of Power Sources 131:340–50

6. Hollmuller P, Joubert J-M, Lachal B, Yvon K (2000) Evaluation of a 5 kW$_p$ photovoltaic hydrogen production and storage installation for a residential home in Switzerland. Int. J. Hydrogen Energy 25:97–109

7. Lehman P A, Chamberlin C E, Pauletto G, Rocheleau M A (1997) Operating experience with a photovoltaic-hydrogen energy system. Int. J. Hydrogen Energy 5 (22):465–470

8. Little M, Thomson M, Infield D G (2005) Control of a DC-interconnected renewable-energy-based stand-alone power supply. Presented at Universities Power Engineering Conference, Cork

9. Little M, Thomson M, Infield D G (2007) Electrical integration of renewable energy into stand-alone power supplies incorporating hydrogen storage. Int. J. Hydrogen Energy 32:1582–1588
10. Szyszka A (1998) Ten years of solar hydrogen demonstration project at Neunburg vorm Wald, Germany. Int. J. Hydrogen Energy 10 (23):849–860

11

Conclusions

Solar hydrogen energy systems are a reliable, independent, sustainable and efficient substitutes to the current centralised energy system based on fossil fuels. The impact of such replacement on the global energy macroeconomy can be potentially enormous. This book does not delve into the political or sociological discussion on such transition but wishes to assert all the same that the decentralisation of energy production and management will affect international politics considerably and will lead to a much safer, cleaner and more enjoyable environment.

From a technological point of view, there are still many developments to be made. For instance, as discussed in the book, hydrogen should be more efficiently stored at ambient temperatures and pressures. But, as already happened many times with other types of technology, a strong demand from the market can generate economies of scale capable of accelerating the development of newer and better solutions.

The efficiencies of the first and the second principles of hydrogen systems are satisfactory for a large-scale deployment, but performance ratios are not the only key success factors of an alternative energy system. In fact when appraising costs and benefits it is extremely important to consider the impact of externalities like pollution reduction, better quality of life, higher gross domestic product, more stable energy supply and energetic independence. The introduction of hydrogen systems can be a breakthrough to solve a myriad of issues from the aspects of human health, politics and energy supply. This is why the implementation of this new energy regime requires a rigorous life cycle analysis, a careful capital budgeting and a meticulous evaluation of the intangible benefits.

Only a correct assessment on the many benefits of solar hydrogen energy systems can foster a deeper understanding of the complete economic picture of this energy alternative. This will alter the way people perceive the environmental and energy issues and profoundly change the governmental energy policies, which will provide a greater thrust for the development of the new energy system and accelerate the transition to a cleaner and better world.

Zini G., Tartarini P.: Solar Hydrogen Energy Systems. Science and Technology for the Hydrogen Economy. DOI 10.1007/978-88-470-1998-0_11, © Springer-Verlag Italia 2012

Subject Index

Library
University of Wisconsin
Fond du Lac